SpringerBriefs in Finance

More information about this series at http://www.springer.com/series/10282

Gagari Chakrabarti · Chitrakalpa Sen

Green Investing

The Case of India

Springer

Gagari Chakrabarti
Department of Economics
Presidency University
Kolkata, West Bengal
India

Chitrakalpa Sen
School of Management
BML Munjal University
Gurgaon, Haryana
India

ISSN 2193-1720 ISSN 2193-1739 (electronic)
ISBN 978-81-322-2025-1 ISBN 978-81-322-2026-8 (eBook)
DOI 10.1007/978-81-322-2026-8

Library of Congress Control Number: 2014946961

Springer New Delhi Heidelberg New York Dordrecht London

Printed on acid-free paper

Springer is part of Springer Science+Business Media (www.springer.com)

Foreword

Thinking Green Technology in a Capitalist Economy

Changes are common, but certain changes are very uncommon. Ecological shift propelled partially by the destructive role of business leaders and corporations is one such very uncommon event that may have far reaching, irreversible, consequences we as humans may not be equipped to deal with. The possibility of extinction of species-being is a scenario far more serious than mere environmental concern that is locally solvable.

In this backdrop, while the matter of difference between social and private costs was well known in economics for a long time, little was done to address the source of the problem. Instead, the problem was displaced to 'solutions' such as emissions trading to incentive producers to produce less pollution. This was like treating the symptom rather than the disease. The universal use of gray technology was hardly affected as a result and the above-mentioned problem of ecological shift could not be turned around. As the gravity of the purported ecological shift becomes clear, recent attempts have shifted gradually toward more basic solutions such as the usage of green technology that are supposed to redress fundamentally the way we produce goods and services. Resultantly, green investment for green technology is one among many long-run solutions being seriously considered now. But it is easier said than done, especially in the context of a globalized capitalist economy. To put the matter bluntly, why should global capitalist firms opening in a competitive market environment substitute gray technologies for green technologies? It cannot be a simple matter of 'consciousness raising' among corporates in this competitive environment where the bottom line is profit. That is, for green technology to have a long-run future in opposition to the current gray technology, it must be profitable for firms to adopt green technology. Green technology in a capitalist economy or green technology with profit is a quite different kind of questioning that the field of economics has only recently begun to think. There is no doubt about the importance of the government's role in research and development of green technology and in proving incentives of various kinds (for example, tax and credit subsidies) for private firms to adopt green technology. However in a competitive market economy, it will ultimately be the connection of profit to adoption of green technology that will hold the key. Isn't the adoption of green technology risky? Is it cost competitive? How does the stock market view green technology vis a vis gray technology? Can it find a place as a fundamental of the corporations or as a variable of the fundamentals of the corporations? Is green technology vulnerable to business cycles, including financial shocks?

Taking off from the premise that green technology and not gray technology is fundamental for our survival, this book picks up on these questions of the relation of green technology to profit maximizing firms in the context of the stock market, competitive environment, and business cycles. It rethinks and invokes various analytical techniques to try to unpack an array of issues associated with green technology. In the topic it deals with and the way it does so, this intervention is a cutting edge contribution to what is definitely becoming one of the most important researchable areas in economics and beyond.

Kolkata, India Anjan Chakrabarti

Contents

About the Authors

Dr. Gagari Chakrabarti completed her Master's, M.Phil., and Doctorate in Economics from the University of Calcutta (India) and is currently working as an Assistant Professor at the prestigious Presidency University (erstwhile Presidency College) in Kolkata, India. Her area of specialization is financial economics including application of econometrics in financial economics. She has several national and international publications to her credit.

Dr. Chitrakalpa Sen is an Assistant Professor in the School of Management, BML Munjal University, Gurgaon, India. An economist by training, Dr. Sen holds an undergraduate degree in Economics and a postgraduate degree in Economics from the University of Calcutta. He did his Ph.D. from the School of Management, West Bengal University of Technology. He has published in a number of national and international peer-reviewed journals, such as *Finance India* and *Journal of Asset Management* and has also authored/co-authored a number of books.

Chapter 1
Prologue

Abstract Global warming has been the biggest threat humankind has faced in the twentieth century. After the industrial revolution, the amount of greenhouse gas in the atmosphere has increased rapidly, leading to a rise in the atmospheric temperature. If greener production technologies are not adapted, by the end of the century, the global climate may change dramatically, causing worldwide catastrophe. Studies have shown that greener companies outperform the other companies even in terms of financial return. Therefore, from the above discussion, it is evident that investment in green and energy-efficient firms is the most profitable choice for wise investors in time to come.

Keywords Global warming · Climate change · Green investment · Sustainable production

> "Global warming, along with the cutting and burning of forests and other critical habitats, is causing the loss of living species at a level comparable to the extinction event that wiped out the dinosaurs 65 million years ago. That event was believed to have been caused by a giant asteroid. This time it is not an asteroid colliding with the Earth and wreaking havoc: it is us." Al Gore, An Inconvenient Truth: The Planetary Emergency of Global Warming and What We Can Do About It.

The latter half of the last century has probably been the most important for humankind. Technological development progressed like never before, at an unprecedented rate; things that were only true in Star Trek movies or science fiction novels were made into reality. Personal computers got smaller and thinner, World Wide Web literally granted the individual access to the information available to the entire world, the advent of cellular phones changed the way the whole world interacts, and many previously life-threatening diseases are eradicated, thanks to the advances in health care. However, basking in this glory, we often fail to recognize a potentially fatal threat that has probably grown bigger at the same rate if not faster as technological advancement, that is the threat of environmental sustainability. The biggest threat to the environment in the last few decades has been that of an ever-increasing carbon signature on earth which led to global

G. Chakrabarti and C. Sen, *Green Investing*, SpringerBriefs in Finance,
DOI 10.1007/978-81-322-2026-8_1

warming. The amount of greenhouse gas on earth has been on an ever-increasing trend. In 2012, the actual greenhouse gas emission has already overshot the target by a massive 1 billion tons.[1] The gap is continually increasing. As a result, the weather is shifting drastically all over the world. In the USA, the summer days are becoming hotter, and in India, monsoon is delayed, East and Southeast Asia is ravaged by recurring hurricanes, and the winter days in the Northern Hemisphere is swept by increasingly intense snowstorms. Without adequate preventive measures, the human-induced damage on the environment will keep on increasing and the earth will reach the tipping point by 2020, when nothing we do will be enough to reverse the faster upward trend in the global temperature (www.UN.org).

At this juncture, it is of particular importance for the firms to make themselves amenable to greener corporate and production policies. A not-so environment-friendly firm imposes significant negative externality on the economy. As a result, the private cost of production and the social cost of production significantly diverge. The excess cost is borne by the society itself. Therefore, from the society's perspective, it is only beneficial to encourage more green technologies. However, from a firm's perspective, a "gray" or environment-unfriendly production technology may prove more profitable. After all, there are not many visionary entrepreneurs who think in really long run and keep sustainability in account. However, with the environmental threat, the awareness is higher than ever. This is slowly but steadily causing a greater tilt toward the greener investment machineries. With increasingly more emphasis on the green technologies, it only makes sense for the firms to go green. Even giants such as Morgan Stanley and Goldman Sachs started owning stakes in green energy projects and set up carbon trading desks (Campbell and Nicholson 2013). Although a tad late, the governments are also gradually realizing the emphasis on green technologies. During the last economic crisis, the G20 countries all went for "green stimulus package" where a proposed 20 % of the fiscal stimulus to be spent on clean technologies.

One common excuse often given by firms is that greener technologies are more expensive and therefore are not profitable to use in the production process. However, with the technological progress, green technologies are getting cheaper day by day and that excuse often sounds ludicrous. In the recent years, the green stocks have come out to be a lucrative investment alternative to investors. With cheaper price tag, green technologies are possible to be made available to public at a more competitive price vis-à-vis the conventional technologies. It increases revenue, creates profit, and captures new market. It is the stuff every firm's dreams are made of. With more and more emphasis on clean and renewable technology, prices of the green stocks are pushing higher and green companies are poised at a better position than its competitors. The green technology era is being compared to the rise of telecommunications in the 1980s and is often dubbed as "mother of all markets." The green technology investment would surge to $226 billion by 2016 (Maxwell 2009) and is poised to surpass even the information technology boom.

[1] *Source* http://digitaljournal.com/article/361682.

Moving to a green and clean technology is inevitability and a structural shift waiting to happen. How soon it happens depends on how soon the importance of green technologies is realized. The discussion so far, some can argue, has been only a normative perspective on why green technologies will thrive, and therefore, more green investments should be made. On a positive side, the importance of green investment can also be established with empirical evidences which will reinforce the arguments above.

Conventionally, the stock valuations of a company are dependent on price-earning ratios, net profits, and debt obligations. However, recent studies show that being energy efficient plays a significant role in pushing the stock prices up. A study by Griffin and Sun (2012) has shown that stock prices rise significantly when a company voluntarily discloses its emission information. The study took 272 companies for 2000–2010 and analyzed their stock price movements from 2 days before to 2 days after the voluntarily emission disclosure. The results suggest that for large companies, this disclosure causes a 0.5 % increase in stock prices, while for the smaller companies, the impact is more profound, about a 2.32 % increase.

The above evidences point toward the fact that investors are increasingly taking energy efficiency into long-term consideration and prefer to invest in an environment-friendly company as low-carbon growth is being considered to be fundamental in determining long-term carbon growth (Carbon Disclosure project 2011). Therefore, it will be only beneficial for the companies to adapt greener and cleaner technologies.

The Carbon Disclosure Project (CDP) conducts emission-based study with 500 largest companies in the world by their market capitalization included in the FTSE Global Equity Index Series, otherwise known as the Global 500. In this study, the companies are sent a questionnaire asking for their carbon strategies. The results from 2013 (CDP 2013) reveal that companies with low-carbon growth, i.e., companies in the Carbon Performance Leadership Index (CPLI), outperform the other companies even in terms of financial return. The result shows that the CPLI companies earn a significantly higher return than the other Global 500 companies and the gap between the returns has significantly increased, especially in the last 2 years.[2]

However, the findings in the CDP 2013 revealed that the big emitters are still not doing enough to reduce their carbon footprint and their emission decisions are often guided by monetary incentives which remain causes of worry.

Therefore, from the above discussion, it is evident that investment in green and energy-efficient firms is the most profitable choice for wise investors in time to come. The alignment between social choice of green technology and the investors' choice of gray technology will be automatically achieved when the green firms are more profitable than the gray ones, in the Indian context. This study tries to answer that very question of investment-worthiness of green instruments. There has been very little research done in this area, especially in the Indian context, and this research

[2] Global 500 Climate Change Report 2013.

attempts to fill in that gap. For that purpose, the authors will develop five different portfolios consisting of 100 % green, 75 % green–25 % gray, 50 % green–50 % gray, 25 % green–75 % gray, and 100 % gray stocks. The research questions are the following:

1. *Do green portfolios possess relatively less own-risk as compared to their gray counterpart?*
2. *How effective the green portfolios are to avoid market risk?*
3. *Are the green portfolios inherently less unstable?*
4. *Do the green portfolios have a higher probability of surviving financial crisis?*
5. *Are the performances of the green backed by their fundamentals?*
6. *Is there any particular technical trading strategy ensuring a consistently above-average return from these portfolios?*

The study will take the following trajectory:

Chapter 2 starts with a review of the green indices available throughout the globe. Then, the study constructs the "pure" green, "pure" gray, and "hybrid" portfolios in Indian market. A stochastic dominance approach is taken in order to compare the own-risk-adjusted return across the portfolios. Next, the study examines any possible volatility and return transmission from the market to the individual portfolios. Apart from the volatility transmission channel, the study also considers the nature and degree of association between individual portfolio returns and the market return using a conditional correlation and empirical survivor function approach. However, the instability may not essentially be external and be internal in nature. For that purpose, any possible presence of deterministic chaos in the five constructed portfolios is examined. The next section delves deeper and examines the survival potential of green instruments during a financial crisis. The last section of Chap. 2 considers individual stocks constituting the green, semi-green, and gray portfolios, rather than the portfolios themselves, and explores possible factors affecting the probability of avoiding crisis for such stocks.

Chapter 3 introduces the concept of trading rules in financial market and reviews some oft-used trading rules. Then, the study makes use of a momentum-based trading rule to examine the investment-worthiness of the green, part green, and gray portfolios.

References

Campbell M, Nicholson CV (2013) Investors seek ways to profit from global warming. Available online at http://www.businessweek.com/articles/2013-03-07/investors-seek-ways-to-profit-from-global-warming. Accessed 16 Dec 2013

Griffin PA, Sun Y (2012) Going green: market reaction to CSR newswire releases. Available at SSRN: http://ssrn.com/abstract=1995132 or http://dx.doi.org/10.2139/ssrn.1995132. Accessed 29 Jan 2012

Maxwell IE (2009) Managing sustainable innovation: the driver for global growth. Springer, Berlin

Chapter 2
Greens—The Obvious Choice Over the Grays?

Abstract This chapter delves into an individual decision-making problem that bears significant social implications. While tagging along less-carbon investment path through increased investment in "green" projects is socially desirable in the modern era, its implementation is not so easy. The policy-makers, however, would sit comfortably if the imperative choice of the new "green" financial products turns out to be, in fact, obvious. This study explores specifically this issue in the context of an emerging market through examining whether given a choice between green and non-green projects, greens become the optimal choice of a rational investor. As is revealed by the study, the green (either completely or partially) portfolios dominate the available alternative gray portfolios. The green portfolios turn out to be the global minimum variance portfolio, and they dominate the gray in terms of the own-risk as well as the market risk. Even the probabilities of surviving crises are higher, and hence, hazard ratios are lower for the green portfolios. Thus, green is preferred to gray and more green is better than less green. Hence, following less-carbon investment path is the most rational and obvious choice for the investors in the Indian market.

Keywords Green investment · Green portfolios · Stochastic dominance · Empirical survivor function · Survival analysis · Hazard ratio

2.1 The Green Indexes

In today's world, the environmental issues in business and investment are gaining increasing significance, particularly in the context of the emerging financial markets. Apart from its visible, adverse environmental impact, growing investment in polluting industries is potentially hazardous due to its associated negative externalities and the resulting market failure that render the efficient market hypothesis and hence the traditional asset-pricing theory useless. The urgency of a drive toward attaining a low-carbon growth path is thus obvious, and this imperative is particularly strong in the context of the emerging markets. Implementation of low-carbon investment strategies, however, requires a proper definition and understanding of

G. Chakrabarti and C. Sen, *Green Investing*, SpringerBriefs in Finance,
DOI 10.1007/978-81-322-2026-8_2

emission landscape across business and its impact on sustainable growth. Some stock markets all over the world have already taken initiatives to ensure a credible information mechanism for the investors through developing "green" indexes where carbon performances can be objectively quantified. Some of these indexes, coming from the developed as well as the emerging market are worth-mentioning.

The European market offers a good number of green indexes, either in isolation or in collaboration with other developed or emerging markets. FTSE's Environmental Market Classification System and Indices provide the world's first comprehensive global classification system for environmental markets. Environmental market companies are defined as providing products and services that deliver solutions to environmental challenges and include environmental technology, also sometimes referred to as "clean tech." The classification system defines environmental market companies and allocates each to the subsector whose definition most closely describes the nature of its business. There are currently six sectors and twenty-four subsectors. The *Low Carbon 100 Europe® Index* is a free-float market capitalization-based index that considers the performance of the hundred largest European companies having the lowest carbon (CO_2) intensity in their respective sectors or homogeneous subsectors. The sustainability index *CEE Responsible Investment Universe* (*CEERIUS*) is a capitalization-weighted price index which is composed of the leading socially committed and ecologically viable companies whose stocks are traded on stock exchanges in the region of Central, Eastern, and Southeastern Europe. The *Euronext FAS IAS®* Index considers those companies whose employees are most represented in share ownership and enables investors, fund managers, and issuers to assess market performances and compare them with those of other listed companies. The *FTSE KLD Global Climate 100 Index* is designed to provide investors with access to investment in the top 100 globally listed companies, whose activities demonstrate the greatest potential for mitigating the immediate and long-term causes of climate change. The *FTSE KLD Global Sustainability* (*GSI*) *Index Series* intends to provide investors with robust index solutions through which they could identify and invest in companies that are committed to long-term environmental, social, and governance sustainability. Various regional sustainability indexes are developed accordingly by considering companies from North America, Europe, and Asia-Pacific regions that are top-ranked in terms of these sustainability criteria. The FTSE Group has collaborated with the Bolsas Mercados Españoles (BME) to introduce the *FTSE4Good IBEX Index.* This index includes those companies from the BME's IBEX 35 Index and the FTSE Spain All Cap Index that meet good standards of practice in corporate social responsibility. These companies seek to ensure sustainable business environment, intend to build up and maintain positive relationships with stakeholders, and endeavor upholding and supporting universal human rights. The *FTSE4Good Index Series* considers four tradable and five benchmark indices, representing global, European, US, Japan (benchmark only), and UK markets. The FTSE4Good benchmark indices include all companies in the broad market index or those that meet the FTSE4Good criteria. *FTSE4Good Environmental Leaders Europe 40 Index* identifies leading European companies with healthy environmental practices. These forty companies belong to the FTSE4Good Index Series and are the top forty companies among those that have obtained the "best practice" environmental rating of 5. The *FTSE4Good Australia 30 Index* intends to provide investors the access to Australian companies that are actively meeting good standards of practice in corporate responsibility. While the FTSE is most active in developing sustainability indexes, there are a few more in the European markets. The *DAXglobal® Alternative Energy Index* is a sector-based global index where investors have the opportunity to invest in the fast-growing and potentially dynamic "alternative energy" sector. The index considers for inclusion in it stocks of only those companies that generate more than 50 % of their revenue in any of the segments of the alternative energy sector such as natural gas, solar power, wind power, ethanol, geothermal, or hydro batteries. The *DAXglobal® Sarasin Sustainability Germany Index* is composed of the one hundred biggest and most liquid German companies based on free-float and market capitalization. Companies are selected according to market capitalization and the average daily trading turnover, and then, they are verified in compliance with the Sarasin Sustainability Matrix®. The *DAXglobal® Sarasin Sustainability Switzerland Index* is composed of the fifty biggest and most liquid Swiss companies based on free-float and market capitalization. The selection of the constituents takes place according to market capitalization and the average daily trading turnover. Thereafter, these companies are verified in compliance with the Sarasin Sustainability Matrix®. Other European markets have adopted similar measures to develop green indexes. The *OMX GES Ethical Index* is one such attempt where the index consisted of all listed companies in Stockholm, Oslo, Helsinki, and Copenhagen, with the exception of those companies that comply with the ethical criteria of the GES Global Ethical Standard and GES Controversial that are based on international standards on environment, human rights, and corruption. Companies with production and/or sales of weapons, tobacco, alcohol, pornography, and gambling are not included. The OMXS30 Ethical Index is ethical version of the OMXS30 Index, and the index family includes OMX GES Ethical Nordic Index, OMX GES Ethical Norway Index, OMX GES Ethical Sweden Index, OMX GES Ethical Denmark Index, OMX GES Ethical Finland Index, and OMX GES OMXS30 Ethical Index. The Austrian stock market has developed a market capitalization-weighted index called the *VBV-Österreichischer Nachhaltigkeits index or the VÖNIX*. The index is comprised of stocks of those Austrian companies, which are best in terms of social and environmental achievements.

The US market has developed a number of indexes to identify clean and sustainable companies. A few of these may be mentioned in this study. The *NASDAQ Clean Edge US Index (CLEN)* is a modified market capitalization-weighted index that considers the best and most active clean energy, publicly traded US companies. The companies included in this index come from the business segments such as manufacturing, development, distribution, and installation of emerging clean energy technologies such as solar photovoltaics, biofuels, and advanced batteries. The five major subsectors that this index encompasses are renewable electricity generation, renewable fuels, energy storage and conversion, energy intelligence, and advanced energy-related materials. The second green US index is the *NASDAQ OMX® Clean Edge® Global Wind Energy Index* which is a modified market capitalization index that is perceived to act as a transparent and liquid benchmark for the global wind energy sector. The constituent companies come from primarily manufacturers, developers, distributors, installers, and users of energy derived from wind sources. The US market also offers an index related to energy-efficient transportation, namely the *Wilder NASDAQ OMX Global Energy Efficient Transport Index* which is a modified, equally weighted index that defines and tracks companies operating globally to develop and promote innovative and energy-efficient modes of transportation. In line with the NASDAQ, the NYSE has also attempted to introduce clean indexes. The *NYSE Arca Environmental Services Index* (*AXENV*) is a modified equal-dollar-weighted index comprised of publicly traded companies that engage in "clean" business activities, trading, and management. Further, the NYSE provides another modified equal-dollar-weighted index, namely *NYSE Arca WilderHill Clean Energy Index* (*ECO*) which is comprised of publicly traded companies whose business stands to benefit substantially from societal transition toward the use of cleaner energy and conservation. The *NYSE ArcaWilderHill Progressive Energy Index* (*WHPRO*) is further provided to consider companies in transition technologies that reduce the carbon or pollutants stemming from coal, oil, and natural gas that enhance efficiency or make efficient utilization of the dominant energy sources. NYSE defines further the "clean tech" sectors as knowledge-based products and services that improve operational performance, productivity, or efficiency while reducing costs, resource and energy consumption, waste, or pollution. The *NYSE ArcaCleantech Index* (*CTIUS*) is a modified equal-dollar-weighted index that takes into account the dominant clean tech companies worldwide.

The Asian emerging markets are equally eager to introduce clean and sustainability indexes. The *SRI-KEHATI Index* has been launched by the Indonesia stock exchange in partnership with KEHATI, the Indonesian Biodiversity Foundation. The companies with assets worth more than US$100 million, a free-float of more than 10 % of the shares, and a positive price-earning ratio are eligible to be included in the index. The companies are evaluated in terms of environment, community involvement, good corporate governance, respect for human rights, business behavior, and labor practices. The Korean market has developed the *Korean SRI Index* that gauges companies' policies, performance, and reporting in terms of environmental sustainability, social commitment, and governance. The SSE and China Securities Index Company Limited have initiated the *SSE Social Responsibility Index* in August 2009 to include one hundred companies that are listed with SSE and perform well in terms of fulfillment of social responsibility. The *Maala SRI* (*Socially Responsible Investing*) *Index* in Tel-Aviv stock exchange includes twenty stocks of "socially responsible" public companies listed in the TA-100 index.

Other emerging stock markets such as the Johannesburg and the Egyptian stock exchanges have started their journey toward green investment. The Egyptian stock exchange (EGX) has signed a memorandum of understanding with the Egyptian Institute of Directors to jointly develop a green index with Standard and Poor's. The Johannesburg stock exchange has launched the *JSE SRI* (*Socially Responsible Investment*) *Index* in 2004 that selects stocks from the FTSE/JSE All Share Index.

The eligible companies must be fit in terms of environmental, societal, and economic sustainability and effective governance. Similarly, the Brazilian stock market offers the *Corporate Sustainability Index* that is basically a portfolio of at most forty stocks that are selected from the stocks traded on the São Paulo stock exchange. While these stocks are most actively traded ones in terms of liquidity, they belong to the companies that are significantly committed to ensure corporate sustainability and social responsibility.

While stock markets all over the world are replacing grays with greens, Indian market has been no exception. Indian market has been among those that are striving to entrench sustainable investment practices in recent years. India has already introduced the S&P BSE GREENEX stock index that constituted of the top twenty-five companies which are good in terms of carbon emissions, free-float market capitalization, and turnover in Bombay stock exchange (BSE). BSE considers the company's initiative to offset the carbon emissions, the offset limit being set to two-third of the company's total emissions. The index is a free-float market capitalization-weighted index comprising of the list of BSE-100 Index. The index has been back-tested from October 1, 2008 (Base Date), with the base index value of 1,000. More green indexes have been launched by the National stock exchange (NSE) in India. The *S&P ESG India Index* provides investors with exposure to a liquid and tradable index of fifty of the best performing stocks in the Indian market as measured by environmental, social, and governance parameters. An Index Committee composed of Standard and Poor's, CRISIL, India Index Services and Products Ltd. (IISL), and KLD maintains the index. The index represents the first of its kind to measure environmental, social, and corporate governing (ESG) practices based on quantitative as opposed to subjective factors. The index employs a unique and innovative methodology that quantifies a company's ESG practices and translates them into a scoring system, which is then used to rank each company against their peers in the Indian market.

Thus, a large number of markets all over the globe have initiated their journey toward achieving green objectives. However, promoting energy-efficient business practices through encouraging investment in these new, environmentally sustainable financial products would require an obvious precondition: Any such "green investment" should be economically viable from the point of view of a participant in the financial market. In cases where following low-carbon investment strategy is optimal for risk-averse players in the market, green ethos could automatically be promoted. This study intervenes particularly in this area to explore whether green investment is an obvious choice over other "non-green," or what this study calls "gray," investments in the context of an emerging financial market, namely India.

2.2 Greens and Grays in the Indian Market

The exploration, whether a "green" investment could win over an equivalent "gray" one, starts from the basic inquiry that whether a "green" portfolio offers the best possible risk-return trade-off to the investors. Moreover, it seeks to explore whether a portfolio with even a slight "green" touch in it wins over its gray counterpart. After finding the optimal green portfolio, we would compare it with other available portfolios in terms of its performance over time, its sensitivity to different economic environment, and its probability of surviving crises.

Toward the purpose, the study considers and compares different portfolios in the Indian market with different extent of "greenness." Specifically, it considers the following portfolios: (a) *GRAY portfolio*: a portfolio that is consisted exclusively of gray stocks, (b) *GREEN portfolio*: a portfolio that is consisted only of green stocks, (c) *G25 portfolio*: a portfolio, 25 % of which is green and the rest is gray, (d) *G50 portfolio*: a portfolio, 50 % of which is green and the rest is gray, and (e) *G75 portfolio*: a portfolio, 75 % of which is green and the rest is gray.

While constructing the 100 % green portfolio, the study uses the stocks constituting the BSE GREENEX, the index for twenty-five green stocks in the Indian market. Incidentally, each green stock is a member of BSE 100, the best-valued hundred stocks in the Indian stock market. While constructing the 100 % gray portfolio, each green stock in the 100 % green portfolio is replaced by a member stock of BSE 200 that is completely gray and comes from the same sector to which the green stock that it replaces belongs to. In construction of a mixed portfolio, say, G25, where 25 % of the total stocks are green, the study retains top 25 % (in terms of risk-adjusted return) of the stocks constituting the 100 % green portfolio and replaces the rest by gray stocks where each green stock is replaced by a gray stock coming from the same sector in BSE 200. The other mixed portfolios are constructed in the same manner.

The green stocks considered in the study are Bharat Heavy Electricals Ltd. (capital goods), Larsen & Toubro Ltd. (capital goods), Titan India Ltd. (consumer durables), HDFC Corp. (finance), ICICI Bank Ltd. (finance), Hindustan Unilever Ltd. (FMCG), ITC Ltd. (FMCG), Cipla Ltd. (health care), Dr. Reddy's Laboratory Ltd. (health care), Lupin Ltd. (health care), DLF Ltd. (housing related), Ultratech Cement Ltd. (housing related), Infosys Ltd. (information technology), Sterlite Industries (India) Ltd. (metal, metal products, and mining), TATA Steel Ltd. (metal, metal products, and mining), GAIL (India) Ltd. (oil and gas), NTPC Ltd. (power), Reliance Infrastructure Ltd. (power), TATA Power Co. Ltd. (power), Bharti Airtel Ltd. (telecom), Bajaj Auto Ltd. (transport equipment), Hero Motocorp Limited (transport equipment), Mahindra & Mahindra Ltd. (transport equipment), Maruti Suzuki India Ltd. (transport equipment), and TATA Motors Ltd. (transport equipment).

The gray stocks considered in the study are members of the BSE 200. We could not incorporate gray stocks from the BSE 100 or BSE SENSEX index as they are all members of either the BSE GREENEX or the BSE Carbonex index. The gray stocks are Havells India Ltd. (capital goods), Tharmax Ltd. (capital goods), Videocon Industries Ltd. (consumer durables), UCO Bank (finance), Indian Overseas Bank (finance), Glaxosmithkline Consumer Healthcare Ltd. (FMCG), Britannia Industries Ltd. (FMCG), Glaxosmithkline Pharmaceuticals Ltd. (health care), Wockhardt Ltd. (health care), Cadila Healthcare Ltd. (health care), India Cements Ltd. (housing related), Madras Cements Ltd. (housing related), Hexaware

Technologies Ltd. (information technology), National Aluminium Co. Ltd. (metal, metal products, and mining), Hindustan Copper Ltd. (metal, metal products, and mining), Gujarat State Petronet Ltd. (oil and gas), CESC (power), Torrent Power Ltd. (power), JSW Energy Ltd. (power), Tata Communications Ltd. (telecom), MRF Ltd. (transport equipment), Amara Raja Batteries Ltd (transport equipment), Eicher Motors Ltd. (transport equipment), Motherson Sumi Systems Ltd. (transport equipment), and Apollo Tyres (transport equipment). Corresponding to each green stock, there is its gray counterpart from the same sector to which the green stock belongs.

The construction of the portfolios is actually a task of assigning weights to different financial assets so as to optimize the investors' objective function. The theory and methodology for assigning portfolio weights are discussed in the next section.

2.2.1 *Construction of Green and Gray Portfolios in Indian Market*

Construction of an efficient portfolio is a standard mean–variance optimization problem. It starts from the premise that the investors participating in the market are risk averse and their preferences can be represented by a (derived) expected utility function defined over the mean and variance of a portfolio's return ($\bar{Z}$, $\bar{\sigma}^2$, respectively). The standard assumption on such preference relation is that investors are induced to prefer higher means and smaller variances (measuring risk). Under this assumption, the group of potentially optimal portfolios for risk-averse investors is hence those with the highest expected return for a given level of variance and simultaneously the smallest variance for a given level of expected return. If short sales are unrestricted, the first condition is a sufficient description (Ingersoll 1987). Such portfolios are called mean–variance efficient portfolios. The study, however, works with a broader class of portfolios, namely the minimum variance portfolios that give the smallest variance at every level of expected return. All mean–variance efficient portfolios are minimum variance portfolios, and the mean–variance analysis is consistent with expected utility maximization. In general, all the minimum variance portfolios are included in the optimal set. Under the assumptions that (a) each individual chooses a portfolio with the objective of maximizing a derived concave utility function of the form $V(\bar{Z}, \bar{\sigma}^2)$ with $V_2 < 0$ and $V_1 > 0$; (b) all investors have a common time horizon and homogeneous belief about $\bar{Z}$ and Σ; (c) each asset is infinitely divisible, and (d) there is a riskless asset that could be bought or sold without any restriction, each investor will hold a minimum variance portfolio. No other portfolio could be optimal because given $V_2 < 0$, the minimum variance portfolio with the same expected return would be preferred to any other alternative.

As they are constructed, all the green, semi-green, and gray portfolios in the Indian stock market are global minimum variance portfolios. The construction requires an analysis of the risk and return of the constituent assets.

2.2.1.1 Risk and Return on the *n*-Asset Portfolio

Consider an n-asset portfolio where R_i is the return on the ith asset ($R_i = \ln P_t/P_{t-1}$), σ_{I} is the standard deviation of return, and X_i is the proportion of wealth invested in the ith asset. The return on the portfolio is then given as follows:

$$R_{\mathrm{PF}} = X'R = (X_1, X_2, \ldots, X_n).\begin{pmatrix} R_1 \\ R_2 \\ \ldots \\ R_n \end{pmatrix} = X_1R_1 + X_2R_2 + \cdots + X_nR_n \quad (2.1)$$

Similarly, the expected return on the portfolio is given as follows:

$$ER_{\mathrm{PF}} = E(X'R) = X'E(R) = X'\mu = (X_1, X_2, \ldots, X_n).\begin{pmatrix} \mu_1 \\ \mu_2 \\ \ldots \\ \mu_n \end{pmatrix} = X_1\mu_1 + X_2\mu_2 + \cdots + X_n\mu_n \quad (2.2)$$

The variance–covariance matrix of return is given by

$$\mathrm{Var}\,(R) = \begin{pmatrix} \mathrm{var}\,(R_1) & \mathrm{cov}\,(R_{12}) & \ldots & \mathrm{cov}\,(R_{1n}) \\ \mathrm{cov}\,(R_{21}) & \mathrm{var}\,(R_2) & \ldots & \mathrm{cov}\,(R_{2n}) \\ \ldots & \ldots & \ldots & \ldots \\ \mathrm{cov}\,(R_{n1}) & \ldots & \ldots & \mathrm{var}\,(R_n) \end{pmatrix} = \Sigma \quad (2.3)$$

The variance–covariance matrix is symmetric making the off-diagonal terms equal, so that $\Sigma = \Sigma'$ (the transpose of Σ).

The variance of the portfolio hence is defined as follows:

$$\begin{aligned} \sigma^2_{(\mathrm{PF})} &= \mathrm{var}\,(X'R) = X'\Sigma \cdot X \\ &= (X_1, X_2, \ldots, X_n) \cdot \begin{pmatrix} \mathrm{var}(R_1) & \mathrm{cov}(R_{12}) & \ldots & \mathrm{cov}(R_{1n}) \\ \mathrm{cov}(R_{21}) & \mathrm{var}(R_2) & \ldots & \mathrm{cov}(R_{2n}) \\ \ldots & \ldots & \ldots & \ldots \\ \mathrm{cov}(R_{n1}) & \ldots & \ldots & \mathrm{var}(R_n) \end{pmatrix} \cdot \begin{pmatrix} X_1 \\ X_2 \\ \ldots \\ X_n \end{pmatrix} \\ &= X_1^2\sigma_1^2 + \cdots + X_n^2\sigma_n^2 + 2\sum_{i,j} X_iX_j\sigma_{ij} \end{aligned} \quad (2.4)$$

where σ_{i}^2 is the variance of R_i and σ_{ij} is the covariance between R_i and R_j ($i \neq j$).

2.2.1.2 Finding the Global Minimum Variance Portfolio

In choosing a global minimum variance portfolio, the risk-averse investor allocates his wealth among the available assets in such a manner that will minimize the risk of investment. In other words, he chooses weights given to different assets in a portfolio so as to make the variance of the chosen portfolio the minimum possible. Hence, the choice of minimum variance portfolio is essentially a choice of weights given to available assets.

The minimum variance portfolio $X = (X_1, X_2, \ldots, X_n)'$ in an n-asset case solves the constrained minimization problem:

$$\min_{X_1, X_2, \ldots, X_n} \sigma_{PF}^2 = X_1^2\sigma_1^2 + \cdots + X_n^2\sigma_n^2 + 2\sum_{i,j} X_i X_j \sigma_{ij}$$

Such that $X_1 + X_2 + \cdots + X_n = 1$

The Lagrangian function for the problem could be set as follows:

$$L(X_1, X_2, \ldots, X_n, \lambda) = X_1^2\sigma_1^2 + \cdots + X_n^2\sigma_n^2 + 2\sum_{i,j} X_i X_j \sigma_{ij} + \lambda(X_1 + X_2 + \cdots + X_n - 1)$$

The first-order conditions are as follows:

$$\frac{\delta L}{\delta \chi_1} = 2 \cdot X_1 \cdot \sigma_1^2 + 2 \cdot X_2 \cdot \sigma_{12} + 2 \cdot X_3 \cdot \sigma_{13} + \cdots + 2 \cdot X_n \cdot \sigma_{1n} + \lambda = 0$$

....

$$\frac{\delta L}{\delta \chi_n} = 2 \cdot X_n \cdot \sigma_n^2 + 2 \cdot X_1 \cdot \sigma_{1n} + 2 \cdot X_2 \cdot \sigma_{2n} + \cdots + 2 \cdot X_{n-1} \cdot \sigma_{n,n-1} + \lambda = 0$$

$$\frac{\delta L}{\delta \chi_n} = X_1 + X_2 + \cdots + X_n - 1 = 0$$

In matrix notation, the minimum variance portfolio choice problem turns out to be

$$\min_{X} \sigma_{PF}^2 = X' \sum X \text{ s.t. } X'1 = 1 \tag{2.5}$$

The first-order conditions are described as follows:

$$\begin{pmatrix} 2.\sigma_1^2 & 2.\sigma_{12} & \ldots & 2.\sigma_{1n} & 1 \\ 2.\sigma_{12} & 2.\sigma_2^2 & \ldots & 2.\sigma_{2n} & 1 \\ \ldots & \ldots & \ldots & \ldots & \ldots \\ 2.\sigma_{1n} & 2.\sigma_{2n} & \ldots & 2.\sigma_n^2 & 1 \\ 1 & 1 & \ldots & 1 & 0 \end{pmatrix} \begin{pmatrix} X_1 \\ X_2 \\ \ldots \\ X_n \\ \lambda \end{pmatrix} = \begin{pmatrix} 0 \\ 0 \\ \ldots \\ 0 \\ 1 \end{pmatrix}$$

$$\text{Or, } \begin{pmatrix} 2\Sigma & 1 \\ 1 & 0 \end{pmatrix} \begin{pmatrix} X \\ \lambda \end{pmatrix} = \begin{pmatrix} 0 \\ 1 \end{pmatrix}$$

Or, this is of the form $A_X \cdot Z_X = b$ where, $\begin{pmatrix} 2\Sigma & 1 \\ 1 & 0 \end{pmatrix} = A_X$, $\begin{pmatrix} X \\ \lambda \end{pmatrix} = Z_X$ and $\begin{pmatrix} 0 \\ 1 \end{pmatrix} = b$

The solution for Z_X will be given by $Z_X = A_X^{-1} \cdot b$. The first n elements in Z_X are the optimal weights that help us construct the global minimum variance portfolio.

2.2.1.3 Portfolio That Gives Maximum Expected Return Subject to Minimum Variance

As mentioned by Markowitz (1951), risk-averse investors would seek to find portfolios that provide the best risk-return trade-off to him. This study constructs the efficient portfolio in two steps. In the first step, the optimal weights are chosen so as to minimize the portfolio return variance, or what is better termed as "risk" of the portfolio. The investors, however, are equally interested in the expected return that the portfolio offers. The best risk-return trade-off would then be offered by a portfolio that gives the maximum expected return subject to a target level of risk which is set to be the minimum variance. Hence, the study in the second step selects the appropriate weights that offer maximum expected return subject to the minimum variance that is obtained through Step 1. The weights derived in the second stage would thus be the final and optimal weights considered in the study.

Thus, in Step 1, the choice problem is $\min_X \sigma^2_{\mathrm{PF}} = X' \sum X$ s.t. $X'1 = 1$ giving $\sigma^2_{\min}$ as the minimum variance. In Step 2, the investor would solve the constrained maximization problem: $\max_X \mu_{\mathrm{PF}} = X'\mu$ s.t. $\sigma^2_{\mathrm{PF}} = X' \sum X = \sigma^2_{\min}$ and $X'1 = 1$. The solution to Step 2, given by X′, would define the "best" portfolio for the investor.

The portfolio members with their associated weights are shown in Table 2.1. Short selling is permitted in our framework.

Using the optimal weights, the minimum variance offered by each portfolio has been calculated along with the associated return. Table 2.2 shows the minimum global variance portfolios with the corresponding risk and return.

The minimum risk that a portfolio could impose (proxied by the return variance) is the lowest for the green portfolio and the highest for its gray counterpart. The green portfolio incidentally offers the highest return and risk-return combination shifts in a favorable direction as one substitutes gray stocks in a portfolio with the green ones. Establishing the green portfolio as the most efficient global minimum variance portfolio is not, however, sufficient to designate it as the investors' optimum choice. Toward the purpose, the study seeks to answer the following set of questions:

1. *Do green portfolios possess relatively less own-risk as compared to their gray counterpart*? Risk-averse investors are interested in risk-adjusted return, specifically the own-risk-adjusted return. Traditionally, the own-risk of any portfolio depends on the risk of the individual security returns quantified by their respective return variances. This study considers the overall portfolio risk, rather than the risks of the individual securities that constitute it. We then check whether the own-risk-adjusted return of green portfolios could dominate that of the gray portfolios.
2. *How effective the green portfolios are to avoid market risk*? Along with the own-risk-adjusted returns, the risk-averse investors are concerned about reducing market or systematic risks of portfolios. In the available literature, market risk of a portfolio depends on the degree of association between the market return and the individual security returns. Our analysis considers market risk of a portfolio in a bit different way. It considers the degree of association between individual portfolio returns and the return on a chosen market benchmark index. It then seeks to explore whether green portfolio return movements are more closely associated with the market movements compared to their gray counterparts. A portfolio whose returns are less affected by changes in market is taken as one with relatively less market risks.

Table 2.1 Green and gray portfolios with their constituent members and weights

Industry	Green portfolio#	75 % Green portfolio	50 % Green portfolio	25 % Green portfolio	Gray portfolio[c]
Capital goods	Bharat Heavy Electricals Ltd.[a]	Havells India Ltd.	Havells India Ltd.	Havells India Ltd.	Havells India Ltd.
	Larsen & Toubro Ltd. [a]	Larsen & Toubro Ltd. #	Larsen & Toubro Ltd. #	Larsen & Toubro Ltd. #	Tharmax Ltd.
Consumer durables	Titan India Ltd. [b]	Titan India Ltd. #	Titan India Ltd. #	Videocon Industries Ltd.	Videocon Industries Ltd.
Finance	HDFC Corp.[a]	HDFC Corp. #	UCO Bank	UCO Bank	UCO Bank
	ICICI Bank Ltd.[a]	ICICI Bank Ltd. #	ICICI Bank Ltd. #	Indian Overseas Bank	Indian Overseas Bank
FMCG	Hindustan Unilever Ltd.[a]	Hindustan Unilever Ltd.#	Glaxosmithkline Consumer Healthcare Ltd.	Glaxosmithkline Consumer Healthcare Ltd.	Glaxosmithkline Consumer Healthcare Ltd.
	ITC Ltd. [a]	ITC Ltd. #	ITC Ltd. #	ITC Ltd. #	Britannia Industries Ltd.
Health care	Cipla Ltd.[b]	Cipla Ltd. #	Cipla Ltd. #	Glaxosmithkline Pharmaceuticals Ltd.	Glaxosmithkline Pharmaceuticals Ltd.
	Dr. Reddy's Laboratory Ltd.[a]	Dr. Reddy's Laboratory Ltd. #	Dr. Reddy's Laboratory Ltd. #	Wockhardt Ltd.	Wockhardt Ltd.
	Lupin Ltd. [b]	Lupin Ltd.#	Lupin Ltd. #	Lupin Ltd. #	Cadila Healthcare Ltd.
Housing related	DLF Ltd. [b]	India Cements Ltd.	India Cements Ltd.	India Cements Ltd.	India Cements Ltd.
	Ultratech Cement Ltd. [b]	Ultratech Cement Ltd. #	Ultratech Cement Ltd. #	Ultratech Cement Ltd. #	Madras Cements Ltd.
Information technology	Infosys Ltd. [a]	Infosys Ltd. #	Hexaware Technologies Ltd.	Hexaware Technologies Ltd.	Hexaware Technologies Ltd.
Metal, metal products, and mining	Sterlite Industries (India) Ltd. [a]	National Aluminium Co. Ltd.	National Aluminium Co. Ltd.	National Aluminium Co. Ltd.	National Aluminium Co. Ltd.
	TATA Steel Ltd. [a]	Hindustan Copper Ltd.	Hindustan Copper Ltd.	Hindustan Copper Ltd.	Hindustan Copper Ltd.

(continued)

Table 2.1 (continued)

Industry	Green portfolio#	75 % Green portfolio	50 % Green portfolio	25 % Green portfolio	Gray portfolio[c]
Oil and gas	GAIL (India) Ltd. [a]	Gujarat State Petronet Ltd.	Gujarat State Petronet Ltd.	Gujarat State Petronet Ltd.	Gujarat State Petronet Ltd.
Power	NTPC Ltd. [a]	CESC	CESC	CESC	CESC
	Reliance Infrastructure Ltd. [a]	Reliance Infrastructure Ltd. #	Torrent Power Ltd.	Torrent Power Ltd.	Torrent Power Ltd.
	TATA Power Co. Ltd.	TATA Power Co. Ltd.#	JSW Energy Ltd.	JSW Energy Ltd.	JSW Energy Ltd.
Telecom	Bharti Airtel Ltd.[a]	Bharti Airtel Ltd. #	Tata Communications Ltd.	Tata Communications Ltd.	Tata Communications Ltd.
Transport equipments	Bajaj Auto Ltd. [a]	Bajaj Auto Ltd. #	Bajaj Auto Ltd. #	MRF Ltd.	MRF Ltd.
	Hero Motocorp Ltd. [a]	Amara Raja Batteries Ltd.	Amara Raja Batteries Ltd.	Amara Raja Batteries Ltd.	Amara Raja Batteries Ltd.
	Mahindra & Mahindra Ltd.[a]	Mahindra & Mahindra Ltd.#	Mahindra & Mahindra Ltd.#	Eicher Motors Ltd.	Eicher Motors Ltd.
	Maruti Suzuki India Ltd.[a]	Maruti Suzuki India Ltd.#	Maruti Suzuki India Ltd.#	Maruti Suzuki India Ltd.#	Motherson Sumi Systems Ltd.,
	TATA Motors Ltd.[a]	TATA Motors Ltd.#	TATA Motors Ltd.#	TATA Motors Ltd.#	Apollo Tyres

Green portfolio weights:−0.0019, 0.0217, 0.0180, 0.0542, −0.0446, 0.0233, 0.1116, 0.0887, 0.1757, 0.1049, −0.0384, 0.0866, 0.0851, −0.0411, 0.0279, 0.0825, 0.1069, −0.0820, 0.0004, 0.0254, 0.0428, 0.0853, 0.0222, 0.0340, 0.0109

G75 portfolio weights: 0.0232, 0.0800, 0.0062, 0.0106, −0.0991, 0.0627, 0.0161, 0.1344, 0.1223, 0.0849, 0.0365, −0.0099, 0.0004, 0.1692, 0.0414, −0.0552, 0.0685, 0.0741, 0.0200, 0.0624, 0.0617, 0.0554, 0.0076, −0.0402, 0.0665

G50 portfolio weights: 0.056, 0.010, 0.011, −0.085, 0.058, 0.058, 0.117, 0.083, 0.159, 0.143, −0.037, 0.040, 0.016, 0.061, 0.009, 0.053, 0.065, 0.041, −0.045, 0.034, 0.058, 0.018, 0.001, −0.018, 0.096

(continued)

Table 2.1 (continued)

Industry	Green portfolio#	75 % Green portfolio	50 % Green portfolio	25 % Green portfolio	Gray portfolio[c]
G25 portfolio weights: 0.000613605, 0.052773914, 0.014876608, 0.100641078, 0.130100716, 0.104338468, 0.011079174, 0.025236487, 0.134312714, 0.029690379, 0.115235674, 0.213266666, 0.026662625,					
−0.002161019, 0.001398199, 0.001195936, −0.01167791, 0.005541701, 0.022564255, 0.00386555, 0.029560992, 0.027003739, −0.033744053, −0.007084391, 0.004708892					
Gray portfolio weights: 0.0161, −0.0040, 0.0251, 0.1016, 0.0642, 0.0283, 0.1346, 0.3912, 0.0342, −0.0024,					
−0.0006, −0.0079, −0.0057, 0.0191, 0.0035, 0.0382, 0.0457, 0.0076, 0.0293, 0.1228, 0.0179, -0.0439, −0.0129, −0.0022					

#Implies that stocks belong to BSE GREENEX
[a]Implies that stocks belong to BSE SENSEX
[b]Implies that stocks belong to BSE 100
[c]Implies that stocks belong to BSE 200

Table 2.2 Expected return and risk on portfolios

Portfolio	Gray	Green	G25	G50	G75
Return	0.000416	0.068885	0.000137	0.011073	0.016976
Variance (MIN)	0.000044	0.000037	0.000041	0.000041	0.000040

3. *Are the green portfolios inherently less unstable*? A modern interdisciplinary school of literature suggests financial markets to be characterized by nonlinear particularly chaotic dynamics. This has significant bearing on the investment decision made by risk-averse investors. A chaotic system is essentially nonlinear and the best way to describe it is as a system that is deterministic but appears random. A chaotic financial market is intrinsically erratic, characterized by no stable equilibrium. Any deviation from the equilibrium will be self-correcting. Hence, volatility will generate endogenously and crashes will be more of a rule rather than aberration. Since a chaotic series cannot be forecasted, policies to smooth out fluctuations are likely to be ineffective. Hence, risk-averse investors will face problems in constructing optimum portfolios in a chaotic market. Endogenously generated volatilities in asset returns and frequent crashes make the financial assets inherently risky. This is particularly where the present study intervenes. It seeks to compare the green, semi-green, and gray portfolios in terms of their intrinsic instability. Specifically, it explores the possible chaotic nature of the constructed portfolios.
4. *Do the green portfolios have a higher probability of surviving financial crisis*? Or, stated alternatively, do they have a lower hazard ratio? While investors consider the return and risk profile of available financial assets, they would be interested in choosing a "shockproof" portfolio or a portfolio that could sustain the financial stress imposed by the cycles of the economy. Investors could be easily induced to follow a less-carbon investment path even if the intrinsically unstable green portfolios could survive economic crises in a more effective way than their gray counterparts. The case for preaching green investment might be even stronger if increased greenness of portfolios could improve the probability of surviving economic crisis. The study seeks to explore the "shockproof" nature of green and gray portfolios through an exploration of the sensitivity and sustainability of these financial assets to the shocks to the system using the traditional sensitivity analysis (where stress comes exogenously) as well as the modern survival analysis (that believes stress to depend on intrinsic vulnerability of a structure).
5. *Are the performances of the green backed by their fundamentals*? Risk-averse investors will be equally interested in the factors influencing the performance of the financial assets that they choose to construct their portfolios. The literature on finance-growth nexus emphasizes the importance of fundamental-backed financial performances of assets as their selection criterion. In the traditional literature, business failure prediction models make use of various statistical techniques in an attempt to estimate the bankruptcy probability of a firm using a set of covariates such as financial ratios and market-related variables. This study considers individual stocks constituting the green, semi-green, and gray portfolios, rather than the portfolios themselves, and explores possible

factors affecting the probability of avoiding crisis for such stocks. The exploration is difficult to be considered at the portfolio level as it would be rather injudicious to define fundamental or financial ratios for portfolios. Hence, the study starts from a firm-level analysis where the probability of avoiding crisis for these firms is anticipated to depend on several covariates. While some of these covariates would be related to the company fundamentals and the market, the rest would reflect the intrinsic nature of the stock concerned.

2.3 Green and the Gray: A Comparative Approach in Terms of Risk and Return

The study now seeks to answer the questions raised earlier in an attempt to compare the performance of green portfolios with that of the gray portfolio.

2.3.1 Comparison of Own-Risk-Adjusted Return for Portfolios: A Stochastic Dominance Approach

Financial asset returns could be compared effectively using the stochastic dominance method. The existence of a dominated financial asset or dominated portfolio is not consistent with the optimizing behavior of investors who are unconstrained in their action and prefer more to less (Ingersoll 1987). The requirements for the existence of dominance, however, are very strict and seldom arise except when very similar assets are considered. This problem could be overcome using the concept of stochastic dominance. Stochastic dominance is an age-old technique to compare or rank different random variables, and the method finds application in determining the efficient sets of non-dominated portfolios and to facilitate the decision-making process. It has been theoretically and empirically used particularly in the fields of finance and risk management, insurance, economics, and statistics over the last four decades. The method is particularly helpful when researchers do not have enough information regarding investors' utilities and the analytical pdfs of random variables. In empirical analysis of portfolio risk management, method of stochastic dominance supplements the classical mean–variance approach that imposes unrealistic assumptions on investors' utility functions and return pdfs. Moreover, the method takes into account higher-order moments, rather than only the first- or second-order moments, and hence incorporates more information regarding the empirical pdfs. Hence, the method finds better application in portfolio management where return distributions are skewed. Moreover, the method of stochastic dominance is consistent with the expected utility theory and it characterizes general preferences of agents. As pointed out by Ingersoll (1987), stochastic dominance does not always require that for one portfolio to dominate another, it should always outperform, at least weakly, the other. For a portfolio to stochastically dominate another, it should not outperform it, but rather its probability of exceeding any given return has to be higher than its competing portfolio. This paper uses the method of stochastic dominance to explore the possible preeminence of the green portfolios over their gray counterparts. The study, however, compares the risk-adjusted portfolio returns rather than the simple returns. This is justified, as risk-averse investors are interested more in risk-adjusted returns.

Before we move on to the empirical analysis, it would be better to consider the theoretical underpinning. We start from a proper definition of risk rather than defining it loosely to be the variance of return series. As pointed out by Ingersoll (1987), if uncertain outcomes x and y have the same expectation, then x is said to be less risky than y for the class of utility functions, U, if no individual with a utility function in U prefers y to x. In other words, $E[u(x)] \geq E[u(y)]$ for all u in U. For some restricted classes, this ordering is complete. For all pairs of random variables x and y with the same mean but different distributions, either x or y is less risky. For the quadratic utility functions, for example, x is less risky than y, if and only if var(x) < var(y). However, even among complete orderings, there are cases (e.g., the cubic utility functions) where variance does not remain the universal measure of riskiness. To define risks in a better way, we resort to the Rothschild–Stiglitz (1970) theorems on risk. As pointed by Rothschild–Stiglitz (1970), *outcome $\tilde{X}$ is weakly less risky than outcome $\tilde{Y}$ if and only if $\tilde{Y}$ is distributed like $\tilde{X}+\bar{\varepsilon}$ and $\bar{\varepsilon}$ is a fair game with respect to $\tilde{X}$*. That is, $\hat{Y} \overset{d}{=} \tilde{X}+\bar{\varepsilon}, E(\bar{\varepsilon}|X) = 0$ for all X. Following Ingersoll (1987), the proof could be recapitulated as follows:

Taking conditional expectations and using Jensen's inequality,

$$E[u(\tilde{X} + \bar{\varepsilon})|X] \leq u(E[u(\tilde{X} + \bar{\varepsilon})|X] = u(E[\bar{X}|X] + E[\bar{\varepsilon}|X]) = u(X)$$

Again, $E[E[u(\tilde{X}+\bar{\varepsilon})|X]] = Eu(\tilde{X}+\bar{\varepsilon}) \leq Eu(\tilde{X})$ and we get $Eu(\tilde{Y}) = Eu(\tilde{X}+\bar{\varepsilon}) \leq Eu(\tilde{X})$.

Thus, the theorem establishes variance as a valid measure of risk for normally distributed outcome. The normal outcome with a larger variance will be, hence, less risky.

Outcomes, however, may not have the same expectations. Ingersoll (1987) is not in favor of using Levy's extension to the Rothschild–Stiglitz definition under such circumstances in the presence of a riskless asset. Ingersoll (1987) considers a more general definition of risk which asserts *that $\tilde{X}$ is (weakly) less risky than $\tilde{Y}$ if there exist constants a and b such that $E(\tilde{X} - a) = E(\tilde{Y} - a)$ and $(\tilde{X} - a)$ is (weakly) less risky than $(\tilde{Y} - b)$* in the sense of Rothschild and Stiglitz. Relative riskiness is invariant of choice of a and b, and the definition could be extended as follows: $\tilde{X}$ is (weakly) less risky than $\tilde{Y}$ if there exist constants a and b if $(\tilde{X} - \bar{X})$ is (weakly) less risky than $(\tilde{Y} - \bar{Y})$. This definition is related to the concept of second-order stochastic dominance. Ingersoll (1987) points out that a random return $\bar{X}$ is weakly (second-order) stochastically dominant over $\bar{Y}$ if $\bar{Y}$ and $\bar{X}+\xi + \bar{\varepsilon}$ have the same distribution with $\xi \leq 0$ and $E[\bar{\varepsilon}|X+\xi] = 0$, where ξ is a non-positive random variable. Second-order dominance results if there is some ξ such that $\bar{X}$ first order stochastically dominates $\bar{X} \geq \xi$ and $\bar{Y}$ is weakly riskier than this latter quantity. The relationship between riskiness and second-order dominance is hence strong: If $\bar{X} \geq \bar{Y}$ and $\bar{Y}$ is weakly riskier than $\bar{X}$, then $\bar{X}$ stochastically dominates $\bar{Y}$

There are two big classes of stochastic dominance tests. The first is based on the inf/sup statistics over the support of the distributions as in McFadden (1989), Klecan et al. (1991), and Kaur et al. (1994). The second class is based on comparison of the distributions over a set of grid points as in Anderson (1996), Dardanoni and Forcina (1998, 1999), and Davidson and Duclos (2000). Stochastic dominance method is used in empirical analysis of portfolio risk management by Kahneman and Tversky (1979), Post and Diltz (1986), Levy (1992, 2006), Gloy et al. (1999), Wirch and Hardy (2001), Kuosmanen (2001, 2004), Giorgi (2002), Post (2003), Giorgi and Post (2004), Meyer et al. (2004), Eeckhoudt et al. (2008), Milos and Chovakec (2008), Hodder et al. (2009), and Sriboonchitta et al. (2010).

In order to check for the supremacy or otherwise of the green portfolios over their gray counterparts, the study explores whether the risk-adjusted returns for green portfolios could stochastically dominate those of the gray portfolios. For each portfolio, the portfolio returns are calculated using the constituent stocks' returns and the optimum weights. The portfolio variance series is estimated using an appropriate GARCH family model that takes into account the presence of non-normality, skewness, fat tail, and volatility clustering in the series.

The GARCH model of order (p, q) is specified by a mean and a variance equation as follows:

Mean equation: $Y_t = X_t\theta + \varepsilon_t$ and

Variance equation: $\sigma_t^2 = \omega + \alpha\varepsilon_{t-1}^2 + \beta\sigma_{t-1}^2$

The mean equation is a function of exogenous variables with an error term. Since σ_t^2 is the one-period ahead forecast variance based on past information, it is called the *conditional variance*. The conditional variance equation is a function of three terms: (i) a constant term, ω, (ii) the news about volatility from the previous period, measured as the lag of the squared residual from the mean equation, ε_{t-1}^2 (the ARCH term), and (iii) the last period's forecast variance, σ_{t-1}^2 (the GARCH term). Often, the financial time series volatility is characterized by leverage effects and asymmetric response toward good and bad news in the market. Simple GARCH model, however, cannot incorporate such asymmetric response of volatility. The study selects an exponential GARCH or EGARCH model on the basis of minimum AIC criterion that could fit the data best. The EGARCH model was proposed by Nelson (1991) where the conditional variance equation is specified as follows:

$$\text{Log}(\sigma_t^2) = \omega + \sum_{j=1}^{q} \beta_j \log(\sigma_{t-j}^2) + \sum_{i=1}^{p} \alpha_i \left| \frac{\varepsilon_{t-i}}{\sigma_{t-i}} \right| + \sum_{k=1}^{r} \gamma_k \frac{\varepsilon_{t-k}}{\sigma_{t-k}} \tag{2.6}$$

There is asymmetric response to volatility (where good and bad news have differential impact on conditional volatility) if γ is statistically significant and leverage effect (where bad news have greater impact on conditional volatility) is present if $\gamma < 0$.

The estimated EGARCH model is used to compute the conditional volatility series of the individual portfolio returns.

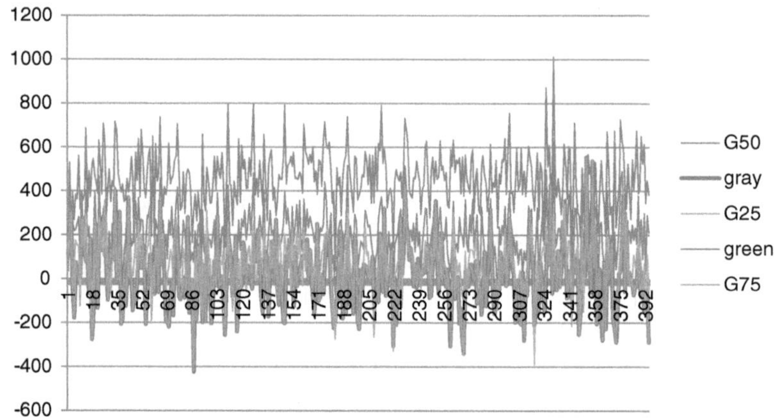

Fig. 2.1 Risk-adjusted return for portfolios

The return (R_t), conditional variance (CV_t), and the 91-day treasury bill rates (RF_t) are used to define portfolio risk-adjusted return (RAR_t) as follows:

$$RAR_t = (R_t - RF_t)/CV_t \tag{2.7}$$

Figure 2.1 depicts the risk-adjusted return series for the five portfolios constructed for the study.

From the diagram, however, it is very difficult to decipher anything on green portfolios' supremacy over their gray counterpart. The testing for possible presence of any stochastic dominance could be appropriate in such cases.

This study follows the methodology of Ng et al. (2011) and tests for the presence of stochastic dominance among portfolio risk-adjusted returns, using a quantile regression technique. Let $x_{11}, x_{12}, \ldots, x_{1n}$ be the series of returns from the first asset and $x_{21}, x_{22}, \ldots, x_{2m}$ be the series of returns from the second asset. Ng et al. (2011) define the two processes as follows:

$$x_{1t} = a_1 + u_t, \quad t = 1, 2, \ldots, n \text{ and}$$
$$x_{2t} = a_2 + v_t, \quad t = n + 1, \ldots, m \text{ with usual properties of } u_t \text{ and } v_t.$$

The cumulative distribution function (CDF) of x_{1t} and x_{2t} will then be given by

$$F(x_{1t}) = P(a_1 + u_t < x) = F_u(x - a_1) \text{ and}$$
$$F(x_{2t}) = P(a_2 + v_t < x) = F_u(x - a_2)$$

Then, the return series of first asset (R_1) will stochastically dominate the return series of second asset (R_2) at first order if $F(x_{2t}) \geq F(x_{1t})$.

As pointed out by Ng et al. (2011), this is equivalent to $F^{-1}(\tau) \geq F^{-1}(\tau)$ or $Q_1(\tau) \geq Q_2(\tau)$.

A dummy variable (D_t) then could be defined that takes the value 1 if $t = 1, 2, \ldots n$ and 0 if $t = n + 1, \ldots, m$.

The combined return data eventually take the following form: $X_t = x_{1t}$, if $t = 1$, 2, … n; and $X_t = x_{2t}$, if $t = n + 1$, …, m. Thus, the model becomes

$$X_t = a + b \cdot D_t + U_t \text{ with } U_t = v_t + (u_t - v_t)D_t; \quad a = a_2 \text{ and } \quad b = a_1 - a_2.$$

The return quantile function is

$$Q_{xt}(\tau|D_t) = a(\tau) + b(\tau) \cdot D_t; \quad \text{where } a(\tau) = a + Q_v(\tau) \text{ and } b(\tau) = b + Q_u(\tau) - Q_v(\tau)$$

Hence,

$$\begin{aligned} Q_{xt}(\tau|D_t = 1) &= a_1 + Q_u(\tau) = Q_1(\tau) \quad \text{and} \\ Q_{xt}(\tau|D_t = 0) &= a_2 + Q_v(\tau) = Q_2(\tau). \end{aligned}$$

And

$$\begin{aligned} b(\tau) &= b + Q_u(\tau) - Q_v(\tau) \\ &= a_1 - a_2 + Q_u(\tau) - Q_v(\tau) \\ &= Q_1(\tau) - Q_2(\tau). \end{aligned}$$

In this framework, the first asset return stochastically dominates the second asset return if

$$\begin{aligned} & Q_{xt}(\tau|D_t = 1) \geq Q_{xt}(\tau|D_t = 0) \\ & \text{or, if } Q_1(\tau) - Q_2(\tau) \geq 0 \\ & \text{or, if, } b(\tau) \geq 0 \end{aligned} \tag{2.8}$$

Hence, first (second) asset return stochastically dominates the second (first) asset return if $\hat{b}(\tau)$ is significantly positive (negative).

The following table (Table 2.3) shows the result of quantile regression on returns of the portfolios constructed for the study. The study selects the portfolios in pairs; that is, it considers any one of the portfolios along with another portfolio

Table 2.3 Stochastic dominance for portfolio risk-adjusted return

Portfolio 1	Portfolio 2	$\hat{b}(\tau)$	Conclusion
Green	Gray	443.78*	Green stochastically dominates gray
G75	Gray	326.14*	G75 stochastically dominates gray
G50	Gray	188.14*	G50 stochastically dominates gray
G25	Gray	59.71*	G25 stochastically dominates gray
Green	G25	384.73*	Green stochastically dominates G25
Green	G50	256.29*	Green stochastically dominates G50
Green	G75	118.29*	Green stochastically dominates G75
G75	G50	138.00*	G75 stochastically dominates G50
G75	G25	266.43*	G75 stochastically dominates G25
G50	G25	128.43*	G50 stochastically dominates G25

* Implies significance at 1 % level

at a time. Hence, we define four pairs between green and gray portfolios: the green and the gray; G25 and gray; G50 and gray; and G75 and gray. In each cases, the green portfolios are taken as the first asset, and thus, for the green portfolios to stochastically dominate the gray portfolio, the coefficient ($\hat{b}(\tau)$) of the dummy variable (that takes the value of 1, if the portfolio is green and 0 otherwise) in the quantile regression should be significantly positive. The stochastic dominance among the green portfolios has also been explored where the study considers green portfolios in six pairs: green and G25; green and G50; green and G75; G25 and G50; G25 and G75; and G50 and G75.

As is revealed by the results, each of the green portfolios stochastically dominates the gray portfolio in terms of risk-adjusted return. Moreover, as is suggested by the coefficient values, dominance over gray portfolio is the maximum for the green portfolio and minimum for the G25 portfolio. Hence, a portfolio with a small number of green stocks in it will of course stochastically dominate a gray portfolio but in a relatively mild way. As the proportion of green stocks increases in the portfolio (from 25 to 50, 75, and 100 %), the dominance becomes stronger. Hence, a portfolio with even a slight touch of green in it will stochastically dominate its gray counterpart in terms of own-risk-adjusted return. Within the green group, once again, the more the extent of greenness, the stronger is the dominance over others. Hence, for the risk-averse investors, green is preferred to gray and more green is better than less green.

Own-risk, however, is not the only consideration to the risk-averse investors, while they choose among portfolios. The non-diversifiable market risks often bother the investors much, and they seek a portfolio that possesses the lowest market risk. This is particularly the area where the study intervenes next and compares the market risk of the five portfolios constructed for the study. The traditional literature computes market risk of a portfolio using individual asset betas. Individual asset beta measures the degree of responsiveness of the individual asset to the movements in the market. The portfolio beta is then calculated as the weighted average of individual betas. This study, however, takes a different approach where the conditional correlation between a portfolio and a benchmark index in Indian market is taken as the proxy for the market risk of the portfolio.

2.3.2 Comparison of Market Risk of Portfolios: Transmission from Market Return and Market Volatility to Individual Portfolios

Investors, who are interested in averting market risk, will tend to consider risks in terms of return as well as volatility. The nature of the market risk of portfolios given by the possible interrelationship between the market and individual portfolios thus may be either in terms of volatility transmission from market to individual portfolios or in terms of interrelationship among the market and individual portfolio returns. The study hence considers the interrelationship between the market index and the constructed green and gray portfolios using two methods. In the first case, we consider how and to what extent the volatility in market gets transmitted to affect the individual green and gray portfolio returns. The nature of the interconnection between market and individual portfolio returns is then analyzed in terms of the conditional correlation between them.

2.3.2.1 Comparison of Market Risk of Portfolios: How Volatility in Market Return Affects the Volatility of Individual Portfolio Returns

The study makes use of a multivariate generalized autoregressive conditional heteroskedasticity (MGARCH) model in its diagonal vector GARCH (VECH) version of Bollerslev et al. (1988). In a diagonal VECH model, the variance–covariance matrix of stock returns is allowed to vary over time and it is more flexible than other multivariate GARCH models. The model is specified (Bollerslev et al. 1988) as follows:

$$\begin{aligned} &\mathrm{VECH}(H_t) = C + \mathrm{AVECH}\left(\Xi_{t-1}\Xi'_{t-1}\right) + \mathrm{BVECH}(H_{t-1}) \\ &\Xi_t|\psi_{t-1} \sim N(0, H_t) \end{aligned} \tag{2.9}$$

H_t is an $n \times n$ conditional variance–covariance matrix. Ξ_t is a 2×1 innovation vector, ψ_{t-1} is the information set at time $t - 1$, C is a $N(N + 1)/2 \times 1$ parameter vector, and A and B are $N(N + 1)/2 \times N(N + 1)/2$ parameter matrices.

As pointed out by Karunanayake et al. (2009), an important issue in estimating a VECH model is the number of parameters to be estimated. To solve the problem, Bollerslev et al. (1988) and Goeij and Marquering (2004) suggested the use of a diagonal form of A and B.

$$h_{ij,t} = \omega_{ij} + \alpha_{ij}u_{i,t-1}u_{j,t-1} + \beta_{ij}h_{ij,t-1} \text{ for } i,j = 1,2 \tag{2.10}$$

A related issue is to ensure the positive semi-definiteness of the variance–covariance matrix. The variance–covariance matrix must always be positive semi-definite. The condition is easily satisfied if all of the parameters in A, B, and C are positive with a positive initial conditional variance–covariance matrix (Bauwens et al. 2006).

The study considers two channels of volatility transmission from market to individual portfolios. Present volatility in individual portfolio return in this study will depend on past news about volatilities in market return, other portfolio returns as well as own return. Parameters in the matrix A will give us idea about such past news or announcement impact on present volatility. A statistically significant a_{ij} will imply cross news effect that gauges the impact of past news impact of volatility in the ith sector on present volatility of the jth sector. Parameters in the matrix B on the other hand will give us idea about past volatility impact on present volatility. A statistically significant b_{ij} will imply cross volatility effect that gauges the impact of past volatility in the ith sector on present volatility of the jth sector. Statistically significant a_{ii} and b_{ii}, however, will give us own past news impact and

Table 2.4 Volatility transmission from market to individual portfolios

Past news impact on present volatility						
Portfolios	G25	G50	G75	Gray	Green	Market
G25	0.0361*					
G50	0.016	0.052**				
G75	−0.001	0.041**	0.023			
Gray	0.025**	0.013	0.017	0.030		
Green	−0.008	0.022	0.017	0.006	0.015	
Market	0.005	0.052	0.041	−0.006	0.034	0.027
Past volatility impact on present volatility						
G25	0.857*					
G50	0.725*	0.578*				
G75	0.750*	0.584*	0.697*			
Gray	0.876*	0.756*	0.756*	0.868*		
Green	0.814*	0.803*	0.803*	0.879*	0.902*	
Market	0.903*	0.890*	0.856*	0.904*	0.792*	0.962*

*Implies significance at 1 % level
**Implies significance at 5 % level

own past volatility impact on present volatility. These measure, respectively, how past news about a sector's own return volatility and past volatility of a sector's own volatility will affect its present return volatility.

The study considers the five portfolios (one gray and four greens of different extent) and the BSE SENSEX (the thirty-stock benchmark index in Indian market) to estimate the multivariate generalized autoregressive conditional heteroskedasticity (MGARCH) model in its diagonal vector GARCH (VECH) version of Bollerslev et al. (1988). The coefficients a_{ij}, a_{ii}, b_{ij}, and b_{ii} are then considered to gauge the volatility impacts. Table 2.4 summarizes the results.

The past news impact of market volatility is not significantly affecting the present return volatilities of individual portfolio returns. G25 and G50 portfolio returns are, however, characterized by significant own past news impact on their present volatility. The analysis of past volatility impact reveals significant observations. Coefficients of all own and cross volatility terms are statistically significant and positive. Hence, present volatility of any portfolio return is significantly and positively affected by past volatility in its own return, other portfolio returns as well as the market returns. Own volatility impact is the strongest in the context of the 100 % green portfolio and the weakest for the G50 portfolio. There has been significant volatility transmission from the market to all the constructed portfolios. As is evident from the values of b_{ii} and b_{ij}, the impact of market has been strongest on the gray portfolio, followed by G75, G50, and G25 portfolios. The impact has been less robust on the 100 % green portfolio. This points out toward relatively less market risk of green portfolios in comparison with their green portfolios. Further, the more the extent of greenness in a portfolio, the lesser the degree of volatility transmission from the market to it.

2.3.2.2 Comparison of Market Risk of Portfolios: A Conditional Correlation and Empirical Survivor Function Approach

Apart from the volatility transmission channel, the study considers the nature and degree of association between individual portfolio returns and the market return. The nature of such comovement over time could be best described using the conditional correlation. Such conditional correlation shows the movement in correlation coefficients between two asset returns over a period of study and hence is useful for assessing how closely the asset returns vary over time with the movements in the market. Single value of correlation coefficient, however, cannot capture the time-varying nature of such movements. Conditional correlations between each of the five portfolio returns and the BSE SENSEX (the thirty-stock benchmark index in Indian market) are estimated using the multivariate generalized autoregressive conditional heteroskedasticity (MGARCH) model in its diagonal vector GARCH (VECH) version of Bollerslev et al. (1988).

Figure 2.2 shows the conditional correlations of different portfolio returns with the market. As is revealed, the conditional correlation between the returns of the gray portfolio and the market is the highest over the study period. The conditional correlation between the 100 % green portfolio return and the market return is significantly lower. However, the relationships of the other green portfolio returns with those of the market are not very clear from these movements in conditional correlations.

To analyze the movements in the conditional correlations in a better way, the study now proceeds to consider the cycles in the conditional correlations using the frequency (band-pass) filter method. The band-pass filter is a linear filter that takes a two-sided weighted moving average of the data where cycles in a "band," given by a specified lower and upper bounds, are "passed" through, or extracted, and the remaining cycles are "filtered" out. To employ a band-pass filter, we first choose the range of durations (periodicities) to pass through. The range is described by a

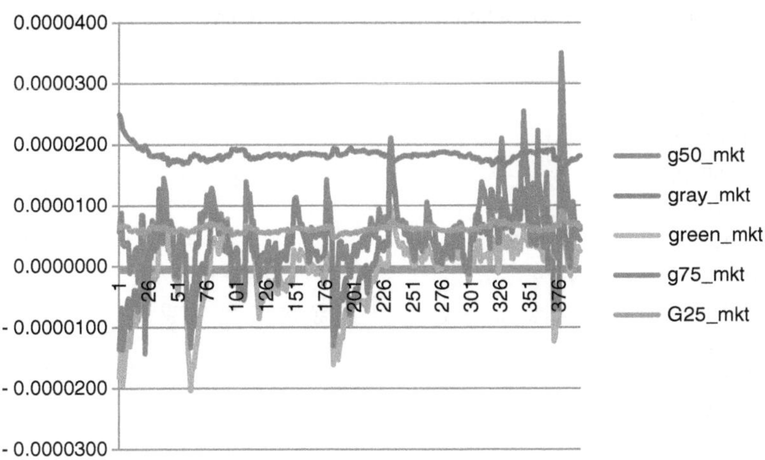

Fig. 2.2 Conditional correlation between portfolio returns and market returns

pair of numbers, specified in units of the frequency of the chosen series. In some cases, researchers may believe that a business cycle lasts somewhere from 1.5 to 8 years so that they intend to extract the cycles in this range. In case the researcher is working with quarterly data, this range corresponds to a low duration of 6 and an upper duration of 32 quarters. Since this study is based on daily data, the lower and upper limits are 540 and 2880, respectively.

The various band-pass filters differ in the way that they compute the moving average. The fixed-length symmetric filters employ a fixed lead/lag length. The symmetric filters are time invariant since the moving average weights depend only on the specified frequency band and do not use the data. There are two forms of this method: One is due to Baxter–King (1999) (BK) and the second to Christiano–Fitzgerald (2003). The two forms differ in the choice of objective function used to select the moving average weights. The full sample asymmetric is the most general filter, where the weights on the leads and lags are allowed to differ. The asymmetric filter is time-varying with the weights both depending on the data and changing for each observation. The study uses this form of filter. The cycles of conditional correlations are shown in the following diagram (Fig. 2.3).

As is evident from the graph of the cycle, the conditional correlations between the market return and the 100 % green portfolio returns are close to zero and remain at a very low level for almost over the entire study period. The cycle for the gray portfolio shows its return to be significantly positively related to that of the market return. As is evident from the cycles of the other mixed portfolios, increases in the extent of greenness in the portfolio are associated with less conditional correlation with market. Hence, the more green the portfolios, the less will be its market risk.

A comparison of the market risks of the different green and gray portfolios could be better performed using the empirical survivor function (Wilson 1927; Brown et al. 2001; Hyndman and Fan 1996) of the conditional correlation series. The empirical survivor function ($S_x(r)$) of a series displays an estimate of the probability of observing a value at least as large as some specified value "r," that is, $S_x(r) = P(x > r) = 1 - F_x(r)$ where $F_x(r)$ is the cumulative density function (CDF).

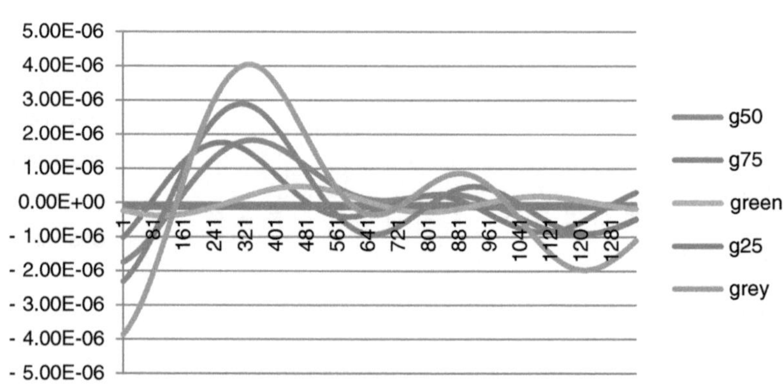

Fig. 2.3 Conditional correlation cycle between portfolio returns and market returns

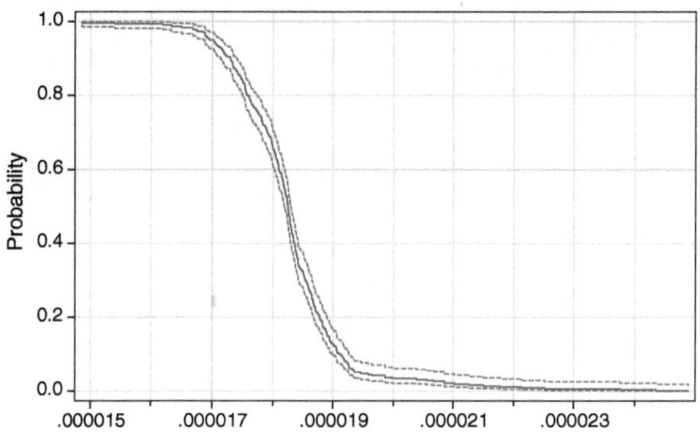

Fig. 2.4 Empirical survivor function for conditional correlation ("gray" and market). *Probability of getting positive conditional correlation: $P(\text{CCOR} > 0) = 1$

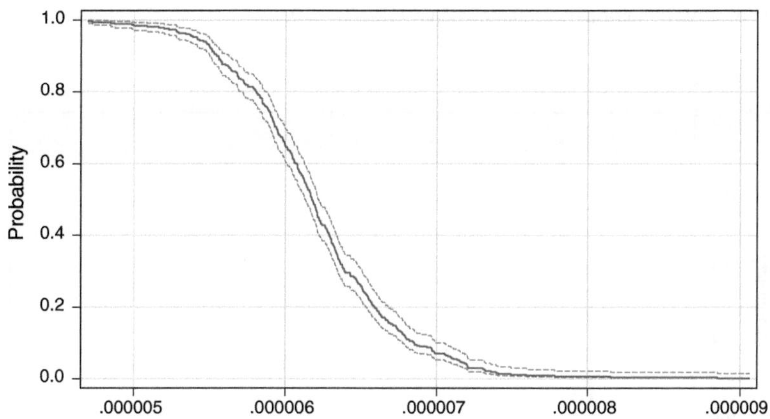

Fig. 2.5 Empirical survivor function for conditional correlation ("G25" and market). *Probability of getting positive conditional correlation: $P(\text{CCOR} > 0) = 1$

The study uses the Rankit method of computing CDF, where given a total number of "n" observations, the CDF for value r is estimated as $(r - ½)/n$. The following figures depict the empirical survivor functions related to different portfolios.

The empirical survivor function in this context gives the probability of getting a specified value of conditional correlation or more. In order to have a lower market risk, the conditional correlation should be either negative or zero. The probability of getting a nonzero, positive conditional correlation should then be lower for a portfolio with lower market risk (Figs. 2.4, 2.5, 2.6, 2.7, and 2.8).

The conditional correlation between the gray portfolio return and the market return has always been positive. As it could be read from the graph, the probability associated with zero conditional correlation happens to be equal to one. Hence, for the gray portfolio return, possibility of getting positive correlation with the market

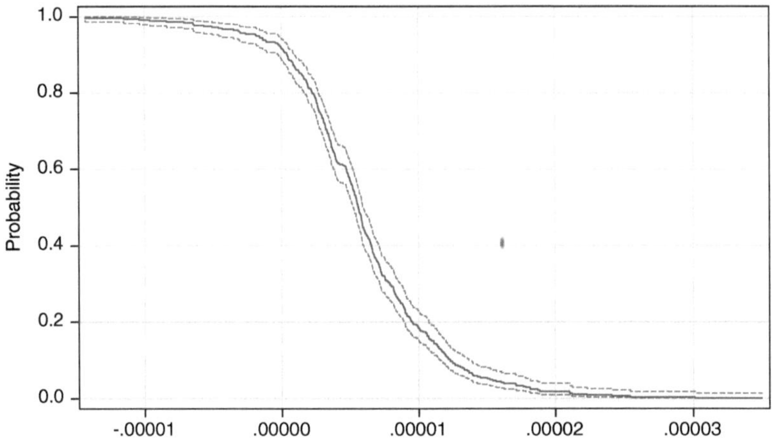

Fig. 2.6 Empirical survivor function for conditional correlation ("G50" and market). *Probability of getting positive conditional correlation: $P(\text{CCOR} > 0) = 0.92$

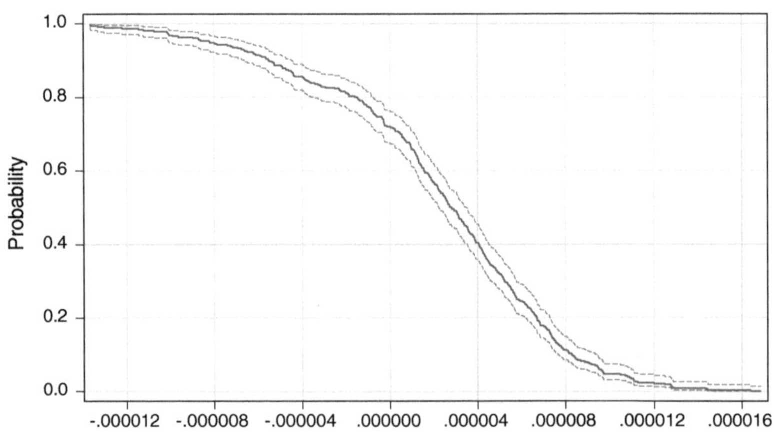

Fig. 2.7 Empirical survivor function for conditional correlation ("G75" and market). *Probability of getting positive conditional correlation: $P(\text{CCOR} > 0) = 0.72$

return is a certain event. This makes the portfolio risky in terms of non-diversifiable market risk. When 25 % of this gray portfolio is replaced by green stock to form the G25 portfolio, probability of getting positive conditional correlation with the market remains at "one," although the values of conditional correlation decrease. Thus, as some gray is replaced by green, the non-diversifiable risk decreases, albeit minimally. The non-diversifiable risk, however, decreases as one increase the extent of "greenness" in the portfolios further. For the G50 and G75 portfolios, probabilities of getting positive conditional correlation are 0.92 and 0.72, respectively. While

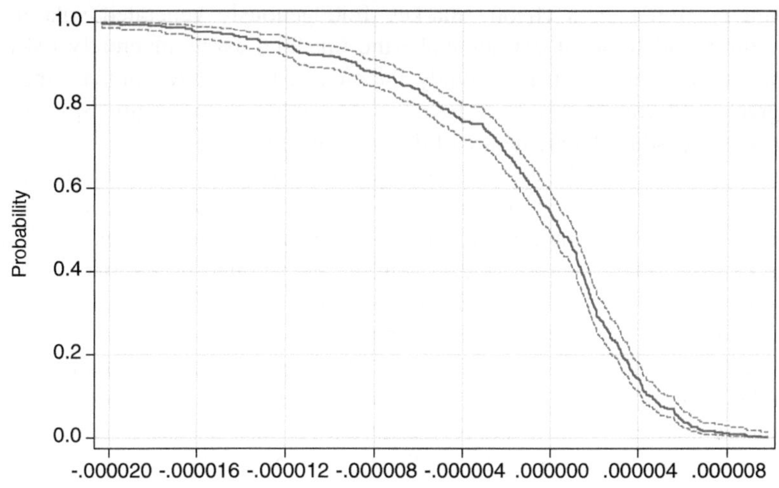

Fig. 2.8 Empirical survivor function for conditional correlation ("green" and market). *Probability of getting positive conditional correlation: $P(\text{CCOR} > 0) = 0.54$

the values of conditional correlation decrease consistently as we move from G50 to G75, G75 offers more negative conditional correlation with the market. The 100 % green portfolio has the minimum non-diversifiable risk in that its returns are mostly negatively correlated with those of the market and have a moderate probability of 0.54 to have positive conditional correlation with market return. Hence, the green portfolios win over their gray counterpart in terms of both the diversifiable and non-diversifiable risks and the greener the portfolio, the lower the risk.

2.4 Are the Green Portfolios Inherently Unstable? A Look into Possible Nonlinearity of Portfolio Returns

As mentioned earlier, one of the significant considerations of investors in a market will be mitigation or at least reduction of the risk associated with the financial asset returns. The source of this risk often lies in the volatility of the return, and such volatility may be exogenous or endogenous in nature. A modern interdisciplinary school of literature suggests financial markets to be characterized by nonlinear particularly chaotic dynamics. This has significant bearing on the investment decision made by risk-averse investors. A chaotic system is essentially nonlinear, and the best way to describe it is as a system that is deterministic but appears random. A chaotic financial market is intrinsically erratic, characterized by no stable equilibrium. Any deviation from the equilibrium will be self-correcting. Hence, volatility will generate endogenously and crashes will be more of a rule rather than aberration. Since a chaotic series cannot be forecasted, policies to smooth out fluctuations are likely to be ineffective. Hence, risk-averse investors will face problems in constructing optimum portfolios in a chaotic market. Endogenously generated volatilities in asset returns and frequent crashes make the financial assets inherently risky. This is particularly where the present study intervenes. It seeks to compare the green, semi-green, and gray portfolios in terms of their intrinsic instability. Specifically, it explores the possible chaotic nature of the constructed portfolios.

For a system to be chaotic, it must have certain specific characteristics. First, it must be *nonlinear*, i.e., in a chaotic system, the time-dependent variables must be related to each other in a nonlinear fashion. Second, it must be *deterministic* in nature, i.e., the future states of the system are determined by the past events, and therefore, although a chaotic system seems random, it is actually rather deterministic. Thirdly and most importantly, a chaotic system should be *sensitive to initial conditions*. The third condition essentially means that very small changes in the initial conditions can build up to significantly large changes in the system's trajectory which can lead to entirely different results. A chaotic system being a nonlinear one, as iterations increase, the error in the system increases exponentially. The error increases so rapidly that after only a small number of iterations, it grows beyond 100 %. Another essential characteristic of a chaotic system is that it continues to evolve with time and two points initially very close to each other but on different trajectories tend to diverge away from each other fairly quickly. Therefore, long-term predictions become very difficult.

If the portfolio return series turns out to be chaotic, it might provide an explanation about the endogenous volatility and instability of the underlying series. Lyapunov exponents in a dynamic system can give an idea about the extent of divergence between two trajectories over time. If the maximum Lyapunov exponent of the system is positive, then deterministic chaos in the system is conclusive. However, in order to find maximum Lyapunov exponent, some essential tests are to be conducted.

2.4.1 Test for Nonlinearity: BDS Test

As discussed earlier, for a system to be chaotic, the first requirement is that it needs to be nonlinear in nature. The BDS test developed by Brock et al. (1987) is used to check for possible nonlinearity in the data. The null hypothesis associated with BDS test is that the data are distributed independently and identically (iid).

The BDS test statistic is given by

$$V_{m,\varepsilon} = \sqrt{T}\frac{C_{m,\varepsilon} - C_{1,\varepsilon}^{m}}{s_{m,\varepsilon}} \tag{2.11}$$

where

$C_{m,\varepsilon}$ correlation integral
m embedding dimension
T time

Table 2.5 Results of BDS test on green and gray portfolio return series

	BDS statistic				
Dimension	G25	G50	G75	Green	Gray
2	0.008383*	0.007827*	0.012955*	0.029614*	0.018280*
3	0.014838*	0.017727*	0.025544*	0.057595*	0.029183*
4	0.021001*	0.022447*	0.031648*	0.076129*	0.033938*
5	0.024897*	0.024670*	0.034854*	0.086863*	0.037621*
6	0.026569*	0.025749*	0.035737*	0.089548*	0.038999*

*Implies significance at 1 % level of significance

$S_{m,\varepsilon}$ standard deviation of $\sqrt{T}(C_{m,\varepsilon} - C^m_{1,\varepsilon})$
$C^m_{1,\varepsilon}$ probability that any two m-dimensional points within a distance ϵ of each other, given x_t are *iid*

$$C^m_{1,\varepsilon} = \Pr(|x_t - x_s| < \varepsilon)^m \tag{2.12}$$

$V_{m,\varepsilon}$ converges in distribution to N(0, 1), i.e., $V_{m,\varepsilon} \xrightarrow{d} N(0,1)$ (Table 2.5).

The results from the above table suggest that the test statistics are all significant at a strong 1 % level, and therefore, the null hypothesis of *iid* is rejected. However, the rejection of the null hypothesis of *iid* is not conclusive to an underlying chaotic behavior. There are four possible scenarios, which can lead to the non-*iid* nature of the data, namely linear dependence, non-stationarity, nonlinear stochastic processes, and nonlinear deterministic process (chaos).

As the portfolio return series is stationary in nature, the linear dependence can be removed by suitable AR filtering. The optimum order of AR is determined by minimum Akaike Information Criterion or AIC (Akaike 1974). However, even after the data are passed through an AR filter, the remaining nonlinearity can come from a stochastic (ARCH-type) model or from chaotic behavior. Therefore, the AR-filtered series is filtered again by a suitable GARCH model. Finally, the AR-GARCH-filtered series is used to test for deterministic chaos.

This study follows the methodology to detect possible chaotic behavior used by Kodba et al. (2004) and Perc (2005). The methodology proposed by Kodba et al. (2004) is explained in the following sections before going into the result and its explanation. This methodology was applied earlier by the same authors in context of other financial markets (Sarkar et al. 2013; Chakrabarti and Sen 2013).

2.4.2 The State Space Reconstruction

First, the state space reconstruction of the data is required as the underlying data are not a state space object (Kantz and Schreiber 2004). The idea behind the state space reconstruction is to replace every state variable with a lagged variable of itself. The vector thus reconstructed will have the same intrinsic characteristics as the original state variables, given a large enough embedding dimension. Using Taken's (1981) embedding theorem, the original system's attractor is reconstructed as follows:

$$p(i) = (y_i, y_{i+\tau}, y_{i+2\tau}, \ldots, y_{i+(m-1)\tau})$$

One advantage of the state space reconstruction is that it can deal with a large dimension and yet can be noise free.

The choice of optimum time delay τ should be very carefully made. The time delay should be large enough to make the set of information contained in y_i and $y_{\tau+i}$ distinguishably different, and at the same time, τ should not be so large that the system does not retain its memory of the initial states. In the next section, the study discusses the methodology behind finding the optimum τ.

2.4.3 Mutual Information Criterion: Finding τ

This study adopts the mutual information criteria (MIC) postulated by Fraser and Sweeney (1986) to calculate the optimum τ. This method measures the dependence between two variables apart by a time delay τ. It is the information available about the state $y_{i+\tau}$ given y_i.

To calculate the MIC, first the observations are arranged in ascending order and t divided in h equal intervals. The MIC is expressed as follows:

$$I(\tau) = -\sum_{h=1}^{j}\sum_{k=1}^{j} P_{h,k}(\tau) \ln \frac{P_{h,k}(\tau)}{P_h P_k} \tag{2.13}$$

where

- P_h probability of the variable falling into the interval h
- P_k probability of the variable falling into the interval k
- $P_{h,k}(\tau)$ joint probability of one variable falling into the interval h and another falling into the interval apart by a time delay τ, given as $h + \tau$.

The optimum time delayτ can be calculated from the first minima of $I(\tau)$. Because, in a chaotic system, as $\tau \to \infty$, $I(\tau) \to 0$ as the correlation between y_h and y_k becomes negligible. At the first minima, $y_{i+\tau}$ adds the most to the available information from y_i without losing the correlation between them completely.

Once the optimum time delay is calculated, the appropriate embedding dimension needs to be determined.

2.4.4 False Nearest Neighborhood: Decide the Optimal m

The optimum embedding dimension is calculated using the method of false nearest neighborhood (FNN) developed by Kennel et al. (1992). According to this model, for an optimum embedding dimension m, the reconstructed delay space has to be topologically consistent with the original state space. Therefore, two points a and b in the "neighborhood," i.e., infinitesimally close to each other, will remain close neighbors if even after a short forward iteration, they do not diverge away from each other. However, if the two points diverge away from each other beyond a particular threshold, they are considered to be *false nearest neighbors* of each other. Choice of the optimum embedding dimension should be made carefully as an embedding dimension too small would cause two points to appear in the neighborhood while actually they are not.

Let $\pi(a)$ be a point on an m-dimensional reconstructed space with a nearest neighbor $p(b)$. If r is the Euclidean distance between two points defined as

$$r(m) = \|y_a(m) - y_b(m)\| \tag{2.14}$$

Then, for nearest neighbors, the distance r is minimized. Next, the system is iterated for a bigger dimension to check whether r is still minimized. The embedding dimension is increased by one so that

$$r(m+1) = \|y_a(m+1) - y_b(m+1)\| \tag{2.15}$$

If $\pi(i)$ is a false nearest neighbor of $\pi(j)$, then $r(m + 1)$ will not be minimized. This is characterized by the change of distance between the two points being larger than an acceptable threshold level when the embedding dimension increased from d to $d + 1$. This can be expressed as follows:

$$\frac{|y_{a+m\tau} - y_{b+m\tau}|}{r(m)} > R_t \tag{2.16}$$

where R_t is the distance ratio threshold. m must be chosen such that the percentage of false nearest neighbors in the data falls to zero. The choice of R_t needs to be made carefully (Rhode and Morari 1997), as too small *a* R_t will not cause the FNN to drop to zero at the correct embedding dimension. And too large *a* R_t tends to accept a lower embedding dimension than actual. However, according to Kennel (1992), $R_t = 10$ proves to be a good choice in most of the cases. However, although the FNN is a widely used process, it is still not robust in the presence of noise.

2.4.5 Determinism Test

Once the embedding delay and the embedding dimension are calculated, the underlying series are tested for determinism, in order to understand whether the data are truly chaotic or a random one that apparently looks like chaotic. Kaplan and Glass (1992) proposed an effective technique to check determinism. Firstly, the attractor is plotted in *a* $x(t)$ versus $x(t - \tau)$ space. Then, the phase space is coarse-grained into $q \times q$ grids. The attractor passes through each grid. A directional vector of unit length, known as the trajectory vector, is assigned to each grid that corresponds to the portion of attractor in it. If e_i be the unit vector passing through each box, then the resultant vector V_k from all the vector passes is just a simple average given by

$$V_k = \frac{1}{P_k} \sum_{i=1}^{P_k} e_i \tag{2.17}$$

where P_k is the number of passes through the kth grid.

If the system is deterministic in nature, the phase space offers a unique solution, i.e., the trajectories inside the grid must never cross. On the other hand for a stochastic system, the trajectories inside the grid cross each other. For a deterministic system, V_k is of unit length, and for stochastic systems, value of V_k is significantly lower than 1.

2.4.6 Maximum Lyapunov Exponent

Lyapunov exponent (Λ) measures the degree of separation between infinitesimally close trajectories in phase space. As discussed earlier, in a chaotic system, the trajectories diverge in time as the system is time dependent and sensitive to initial conditions. For an m-dimensional system, there will be m different Lyapunov exponents. However, the most important would be the maximum Lyapunov exponent ($\Lambda_{\max}$). For *a* $\Lambda_{\max} > 0$, the system would be chaotic and the close trajectories will eventually diverge in state space.

This study uses the method of calculation of $\Lambda_{\max}$ proposed by Wolf et al. (1985). First, an initial point $\pi(0)$ is chosen with a nearest neighbor, a point very close to it. The distance between these two points is considered to be D_0. Next, the two points are iterated forward by time t_{evolv} (equal to τ) and the distance after the iteration is noted (D_{evolv}). If $L_{\text{evolv}} > L_0$, then the system is chaotic as the trajectories diverge in time. The value of t_{evolv} has to be less than $m\tau$, as a larger value will underestimate the value of $\Lambda_{\max}$. At the end of first iteration, a replacement is done to choose a new nearest neighbor for the evolved $\pi(0)$. This process continues till $\pi(0)$ reaches the end of the series. Maximum Lyapunov exponent can be presented as follows:

$$\Lambda_{\max} = \frac{1}{M t_{\text{evolv}}} \sum_{i=0}^{M} \ln \frac{D_{\text{evolv}}^{(i)}}{D_0^{(i)}} \tag{2.18}$$

The results found using the above methodology are summarized below (Table 2.6).

The exploration into possible chaotic nature of portfolio returns reveals interesting results. All the portfolio returns are deterministic among which the green and gray portfolio returns are chaotic with the green portfolio return series being more chaotic than its gray counterpart. The semi-green portfolios, however, are non-chaotic in nature. Hence, the green and the gray portfolio returns are inherently unstable, and volatility is endogenous to these systems. This may disappoint the risk-averse investors who tend to choose between hundred percent green and hundred percent grays. The market, however, offers some scope for diversification. Given the fact that semi-green portfolios are non-chaotic, a proper combination of greens and grays could help investors avoid the intrinsic risk of investment in the market.

Table 2.6 Detection of possible chaos in green and gray portfolio return series

	Portfolios				
	Green	G25	G50	G75	Gray
τ (Emb. delay)	1	1	1	1	3
Shannon entropy	2.13	3.75	3.71	3.79	3.66
m (Emb. dimension)	6	4	4	3	3
Determinism	0.6003	0.5255	0.8833	0.8713	0.6703
Maximum Lyapunov	0.4002	−1.1037	−0.9045	−0.1622	0.3440

While portfolio returns particularly the green returns are chaotic and hence inherently unstable, it would be very difficult for risk-averse investors to protect themselves against volatility that generates endogenously in a financial market. As discussed earlier, in a chaotic financial market, cycles and crashes will be ruled rather than aberration. Therefore, apart from choosing a portfolio that gives lower systematic and non-systematic risks, any rational investor will seek to pick up one that could sustain financial stress imposed by the cycles of the economy. Investors could easily be induced to follow a less-carbon investment path even if the intrinsically unstable green portfolios could survive economic crises in a more effective way than their gray counterparts. The case for preaching green investment might be even stronger if increased greenness of portfolios could improve the probability of surviving economic crisis.

2.5 How Shockproof the Green Portfolios Are: A Survival Analysis

In traditional scenario or sensitivity portfolio analyses, either the analysts start from a subjective definition of optimistic/pessimistic states or the scenarios are exogenously given to them. Portfolio performances over the different scenarios or states of nature are then compared to explore the possible sensitivity and sustainability of these financial assets to the shocks to the system. There is an alternative school of thought that deviates from this traditional sensitivity analysis in that it neither considers financial crises to be exogenously given nor does it define crisis subjectively. Rather, it believes that stress depends on intrinsic vulnerability of a structure and it is a force exerted on by uncertainty and changing expectations of loss in financial markets. It is a continuous variable with a spectrum of values, where extreme values are called financial crises (Illing and Liu 2003). Hence, it concedes that it must be the performance of the asset itself, from where information should be extracted regarding the potential stresses that this asset might face, the asset's susceptibility to such stresses and its potential to survive or endure them. This study analyzes the potential for green portfolios to avoid financial stress following the traditional sensitivity analysis as well as the alternative school of survival analysis.

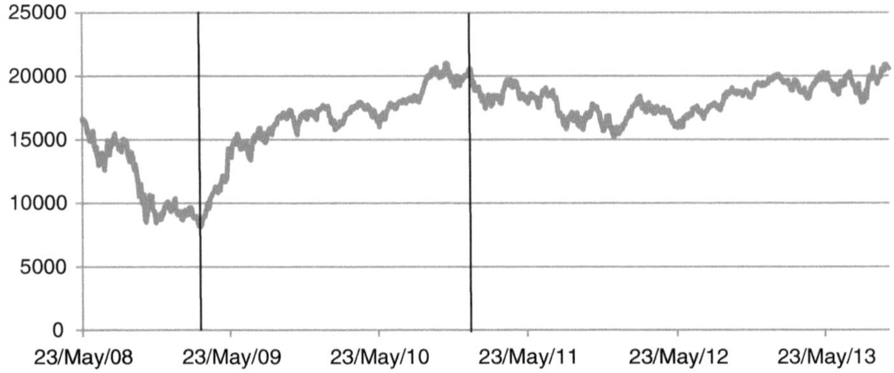

Fig. 2.9 Movements in BSE SENSEX in recent years

2.5.1 Potential of the Green Portfolios to Survive Financial Crisis: A Scenario Analysis

In a scenario analysis, some crisis periods are exogenously chosen to explore the sensitivity or otherwise of different financial assets to such crises. This study selects the recent global financial meltdown of 2007–2008 and its aftermath as the two scenarios. The two scenarios are chosen on the basis of stock price movements in the Indian market. Figure 2.9 shows the movements in BSE SENSEX over the recent years. Starting from May 2008, the market suffered a crisis until March 2009. There has been a recovery in the market since then that continued till December 2010. Hence, the study considers the following two scenarios: (i) scenario I: the period of turmoil or crisis and (ii) scenario II: the period of recovery. The period that immediately follows scenario II is the period of analysis chosen by this study. Scenario I starts from May 2008 (and unfortunately misses the full essence of the crisis that initiated in India in January 2008) as the BSE GREENEX does not have data prior to May 2008.

The performance of the constructed five (one gray, one green, and three mixed) portfolios is examined and compared over these scenarios on the basis of returns as well as risks. The methodology is simple. Using the past price data, portfolio return and risk (given by variance) are constructed for each of these scenarios. The weights given to green and gray stocks to construct portfolios are the same as those used in the current period. Changing portfolio weights will not be commensurate with the traditional scenario analysis.

2.5.1.1 Performance of Green, Semi-green, and Gray Portfolios in Scenario I

The risk-adjusted return series are constructed using the portfolio returns, conditional variability, and risk-free treasury bill rates. Figure 2.10 shows the movements in risk-adjusted return in scenario I. However, the figure is not sufficient to conclude anything regarding the dominance of green portfolios over the gray one.

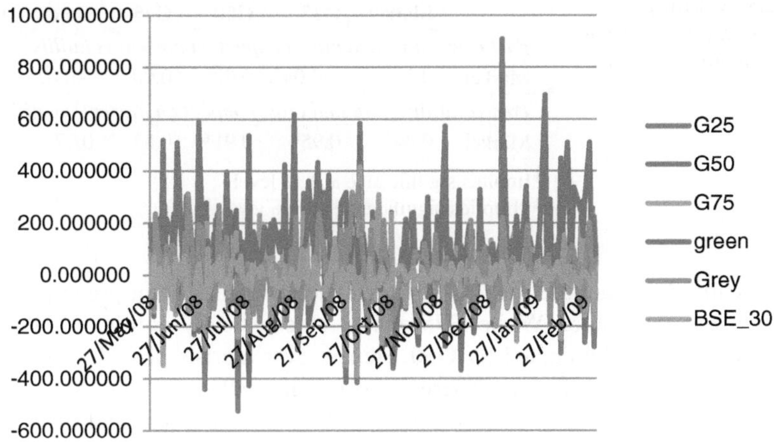

Fig. 2.10 Movements in portfolio risk-adjusted return—scenario I

The study is then extended to explore the possible stochastic dominance of one portfolio return over the others. The results of running quartile regression (using the methodology of Sect. 2.3.1) are shown in Table 2.7.

Over the crisis period, the green portfolio returns stochastically dominate the returns of the gray and other semi-green portfolios. The dominance over the gray has been the maximum, as is evident from the values of $\hat{b}(\tau)$ followed by the dominance over G75, G50, and G25. Thus, during a crisis, it is the 100 % green portfolio that dominates all others. However, a portfolio which is less than 100 % green is not able to outperform the gray or even the other semi-green portfolios. This is evident from the result that none of the G25, G50, or G75 portfolios could

Table 2.7 Stochastic dominance for portfolio risk-adjusted return—scenario I

Portfolio 1	Portfolio 2	$\hat{b}(\tau)$	Conclusion
100 % green over the rest			
Green	Gray	85.88*	Green stochastically dominates gray
Green	G25	84.01*	Green stochastically dominates G25
Green	G50	67.20*	Green stochastically dominates G50
Green	G75	66.73*	Green stochastically dominates G75
Semi-green over the gray			
G75	Gray	−1.79	G75 does not stochastically dominate gray
G50	Gray	16.12	G50 does not stochastically dominate gray
G25	Gray	18.14	G25 does not stochastically dominate gray
Semi-green over the other semi-greens			
G75	G50	−18.51	G75 does not stochastically dominate G50
G75	G25	−18.99	G75 does not stochastically dominate G25
G50	G25	2.08	G50 does not stochastically dominate G25

*Implies significance at 1 % level

Table 2.8 Volatility transmission from market to portfolio returns—scenario I

	Market	G25	G50	G75	Green	Gray
Past news (own and cross) impact on present volatility						
Market	0.10**	−0.04	0.02	0.00	−0.07	0.11**
Past volatility (own and cross) impact on present volatility						
Market	0.84*	0.95*	0.91*	0.83*	0.77	0.80*

*Implies significance at 1 % level
**Implies significance at 5 % level

stochastically dominate the gray. Given a choice among all semi-green portfolios, none is better than other. While G75 cannot stochastically dominate either G50 or G25, G25 is not dominated by G50. Hence, over a crisis period, it is a choice between all-green and all-gray. There is no middle path to follow. Any portfolio that mixes green with gray will be dominated by the all-green, and these cannot dominate the gray. Moreover, an increase in the extent of greenness is not helpful for the investors to fetch significantly higher risk-adjusted return. Hence, it is the 100 % green portfolio that could sustain the exogenously given financial crises in the economy.

The degree and extent of comovement between green and gray portfolio returns with that of the market in scenario I could now be analyzed in terms of volatility transmission and nature of conditional correlations. Estimation of suitable MVGARCH model (following methodology of Sect. 2.3.2) reveals the results depicted in Table 2.8.

So far as the news impact on present volatility is concerned, past news about market volatility is significantly and positively affecting the present volatility in the gray portfolio return and in the market itself. The effect on gray portfolio is more or less similar to that on the market itself. Hence, the gray closely resembles the market. Other green portfolios are not at all affected by past news impact in the market. The past volatility impacts on present volatility, however, are stronger than the past news impact, and such impacts are significantly positive for all portfolios except the 100 % green portfolio. Hence, past volatility in the market is significantly increasing present volatility in the gray and semi-green portfolios. As is revealed by the b_{ij} coefficients, the semi-green portfolios are worse affected than the gray one. Therefore, in a crisis period, it is the green that should be the optimum choice of the risk-averse investor. While other semi-green portfolios are not affected by past news impact from the market, these portfolio returns became excessively volatile when the market had remained volatile in the past. The result reinforces the previous finding. There are significant volatility transmissions from the market to the gray and semi-green portfolios in a crisis period such as that depicted by scenario I. Hence, it pays only when investors opt for hundred percent green portfolios. Mixing of green with gray, however, is not profitable.

The results are reaffirmed when we resort to an analysis of the conditional correlation between market and individual portfolio returns. This section makes use of the empirical survivor function following the methodology of Sect. 2.3.2.2. The empirical survivor functions are depicted in the following Fig. 2.11.

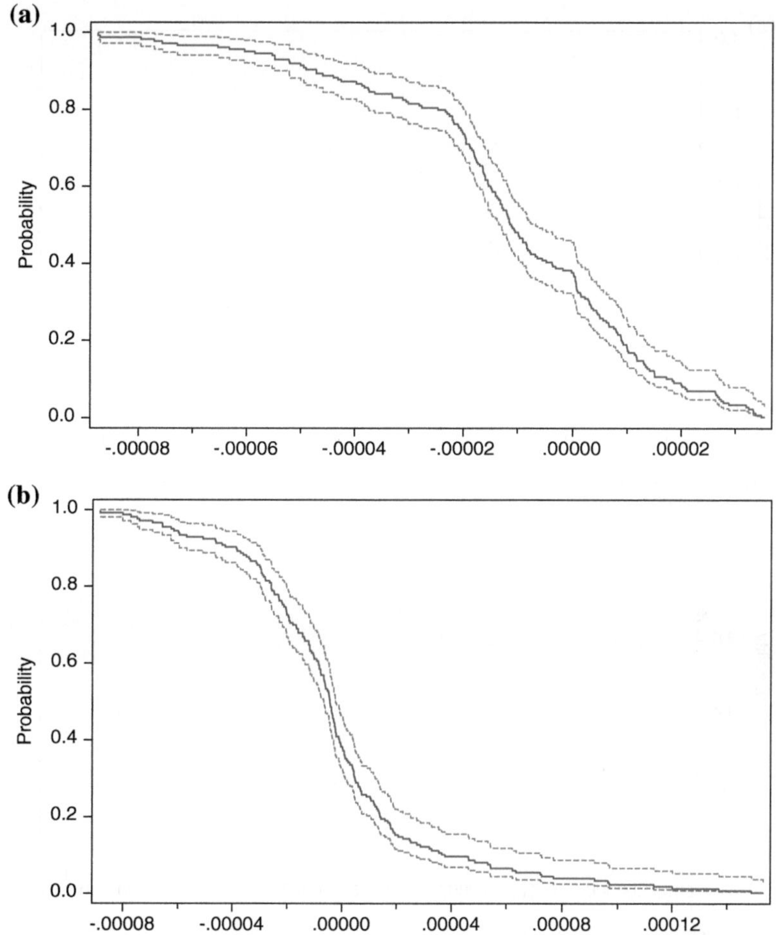

Fig. 2.11 Empirical survivor function for conditional correlation. **a** Gray and market [prob (CCOR > 0) = 0.39]. **b** G25 and market [prob (CCOR > 0) = 0.38]. **c** G50 and market [prob (CCOR > 0) = 0.38]. **d** G75 and market [prob (CCOR > 0) = 0.03]. **e** Green and market [prob (CCOR > 0) = 0.00]

The empirical survivor function shows negative correlation between market and gray portfolio returns for a sufficiently long range. There is, however, a range where this correlation turns out to be positive. The probability of getting positive conditional correlation between market return and gray portfolio return remains at 0.39.

G25 portfolio returns are significantly positively correlated with market return for a wide range of probabilities. The probability of getting strictly positive conditional correlation with market is 0.38 which is slightly lower than the probability of getting positive conditional correlation between gray return and the market return.

Like the G25 portfolio, G50 portfolio returns are significantly positively correlated with market return for a wide range of probabilities. The probability

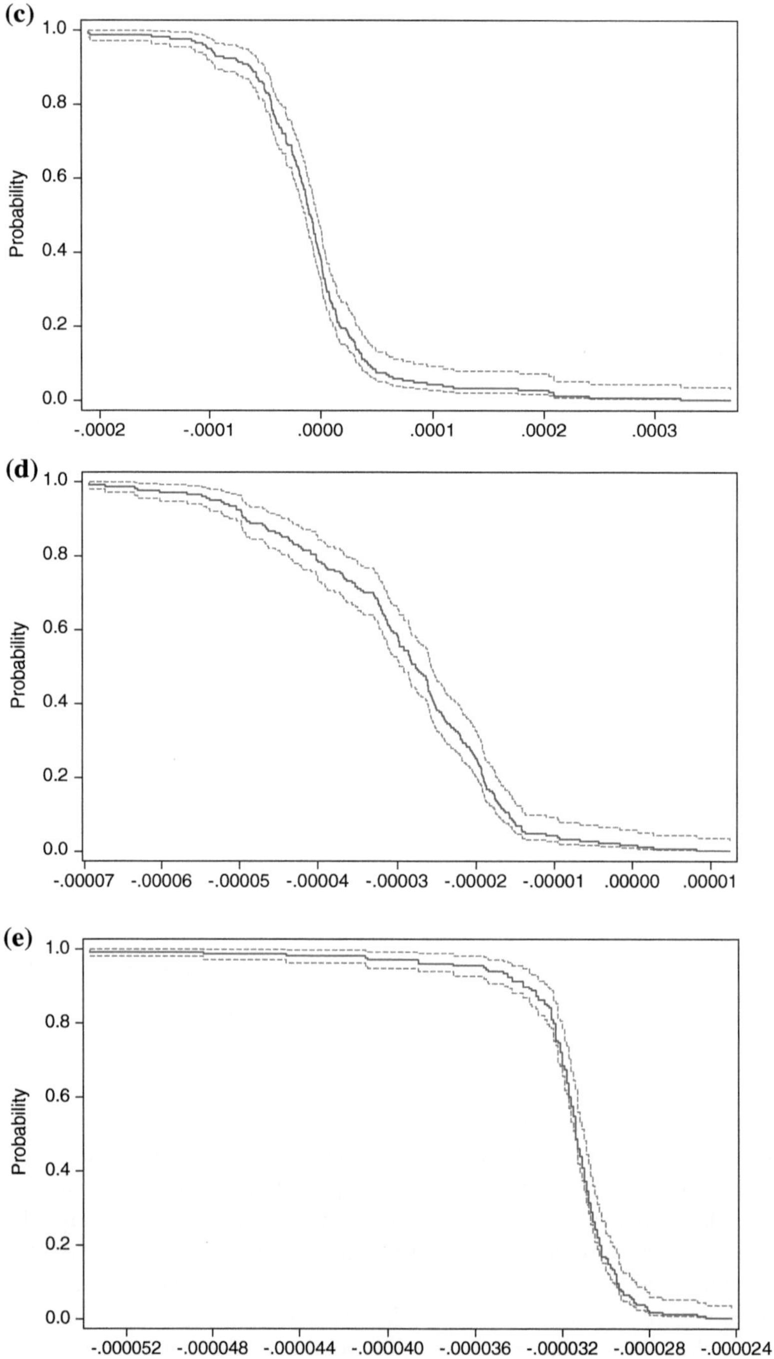

Fig. 2.11 (continued)

of getting strictly positive conditional correlation with market is 0.38 which is slightly lower than the probability of getting positive conditional correlation between gray return and the market return and is similar to that obtained for G25 portfolio.

Unlike the G25 and G50 portfolios, G75 portfolio returns are significantly negatively correlated with market return for a wide range of probabilities. The probability of getting strictly positive conditional correlation with market is only 0.03 which is significantly lower than the probability of getting positive conditional correlation between the market return and the individual gray and other semi-green portfolio returns.

Unlike the gray and the other semi-green portfolios, 100 % green portfolio returns are always negatively correlated with market return. The fact that the green portfolio returns are never positively associated with the market return makes it least vulnerable with respect to the market movements. The green portfolio returns are always moving in the opposite direction of the market even during a period of crisis, thus offering maximum possibility of and gains from diversification.

Thus, in scenario I which has been the period of global financial meltdown, the hundred percent green portfolio becomes the obvious choice of the risk-averse investors. This portfolio dominates all the other gray and semi-green portfolios in terms of risk-adjusted return and gains from diversification. However, the choice remains only between green and gray; semi-green portfolios cannot win over the gray.

The study now proceeds to consider the performances of the constructed portfolios over scenario II. The period, extended from April 2009 to December 2010, is the period of recovery in the market. The following section replicates the methodologies of Sect. 5.1.2 and reports and compares the results.

2.5.1.2 Performance of Green, Semi-green, and Gray Portfolios in Scenario II

The movements in the risk-adjusted returns of the constructed portfolios in scenario II are depicted in Fig. 2.12.

The market offers average risk-adjusted returns, while G75 and green portfolios offer highest risk-adjusted returns. Gray returns are comparatively lower. However, the figure is not sufficient to compare the portfolio returns properly. That is why this section replicates the other methods of comparing portfolio performances that have been used in the earlier section to delve into the optimum choice of the risk-averse investor in scenario II.

Table 2.9 shows the results of quantile regression for scenario II to explore whether portfolios could possibly stochastically dominate one another in terms of their risk-adjusted returns.

The period of recovery offers an interesting scope for diversification. While all the green and semi-green portfolios dominate the gray, green portfolio cannot dominate (and nor is dominated by) the two other semi-green portfolios, namely the G25 and G50 portfolios. The G75 portfolio that is constituted of 75 % of best green stocks and 25 % of best gray stocks performs best in scenario II. G75 dominates all

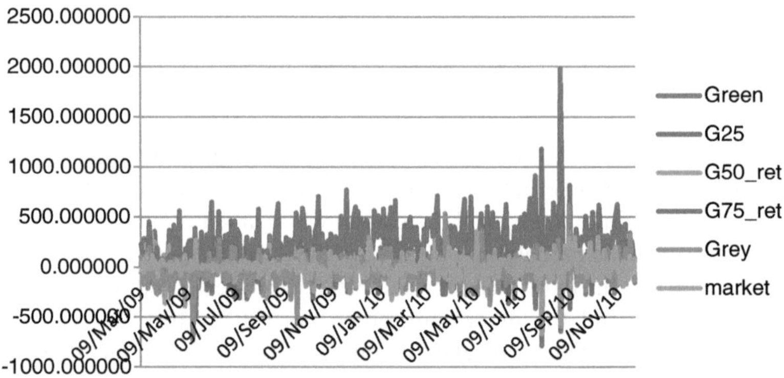

Fig. 2.12 Movements in portfolio risk-adjusted return—scenario II

Table 2.9 Stochastic dominance for portfolio risk-adjusted return—scenario II

Portfolio 1	Portfolio 2	$\hat{b}(\tau)$	Conclusion
100 % green over the rest			
Green	Gray	233.87*	Green stochastically dominates gray
Green	G25	−0.36	Green does not stochastically dominate G25
Green	G50	9.63	Green does not stochastically dominate G50
Green	G75	−199.72	Green is stochastically dominated by G75
Semi-green over the gray			
G75	Gray	40.48*	G75 stochastically dominates gray
G50	Gray	28.30*	G50 stochastically dominates gray
G25	Gray	35.78*	G25 stochastically dominates gray
Semi-green over the other semi-greens			
G75	G50	208.48*	G75 stochastically dominates G50
G75	G25	200.82*	G75 stochastically dominates G25
G50	G25	−7.66	G50 does not stochastically dominates G25

*Implies significance at 1 % level

the other semi-green, the gray, and even the hundred percent green portfolio. Thus, in scenario II, green becomes the obvious choice. But the investors have more profitable opportunity for diversification. A completely green portfolio is dominated by another, which is constructed by a combination of gray and green. The choice of stocks in the optimum portfolio, however, is biased toward green.

The degree and extent of comovement between green and gray portfolio returns with that of the market in scenario II could now be analyzed in terms of volatility transmission and nature of conditional correlations. Estimation of suitable MVGARCH model (following methodology of Sect. 2.3.2) reveals the results depicted in Table 2.10.

The nature of volatility transmission in the recovery period is different from what we observed for scenario I. In scenario II, there has been no past news

Table 2.10 Volatility transmission from market to portfolio returns—scenario II

	G25	G50	G75	Green	Gray	Market
Past news (own and cross) impact on present volatility						
Market	0.014	−0.013	0.003	0.027	−0.035	−0.0005
Past volatility (own and cross) impact on present volatility						
Market	0.90*	0.70	0.65	0.07	0.97*	1.00*

*Implies significance at 1 % level
**Implies significance at 5 % level

impact about market volatility on the green and semi-green portfolios. The nature of the effect of past volatility on present volatility, however, is different. Present level of the gray portfolio volatility is significantly and positively related to past volatility in the market. That is, a volatile market will significantly transmit its volatility to the gray portfolio. Similarly, the G25 portfolio that contains the 25 % green stocks is affected by market volatility. Hence, during the recovery period, the green and semi-greens become the obvious choice of any risk-averse investor. So far as the semi-green portfolios are concerned, portfolios containing relatively more green (such as G50 or G75) than G25 are completely decoupled of market movements.

The results are once again validated when we resort to an analysis of the conditional correlation between market and individual portfolio returns. This section makes use of the empirical survivor function following the methodology of Sect. 2.3.3. The empirical survivor functions are depicted in the following Fig. 2.13.

In scenario II (the recovery period), the gray and green portfolios have shown positive and negative conditional correlation with market returns. The green portfolio offers the lowest probability of having positive conditional correlation, followed by G75, G50, and G25. The gray portfolio shows the maximum probability of having positive conditional correlation with the market.

The scenario analysis thus reveals interesting features of green, semi-green, and gray portfolios. Green portfolio establishes itself as the obvious choice of the investors as it could significantly avoid the market risks generated by the cycles of the economy. Semi-green portfolios can win over the gray only when the economy pulls it out of the recession. However, during recovery, portfolio with very little amount of green stock in it (G25) cannot dominate the gray portfolio in the sense that both are strongly associated with the market. But given a choice between G25 and gray, the investors would favor the G25 as it is less closely associated with the market compared to the 100 % gray portfolio. The gray portfolio thus has always been dominated by the green and occasionally by the semi-green portfolios. Hence, the following less-carbon investment path becomes the obvious choice of the investors in the Indian market.

In traditional scenario or sensitivity analyses, either the analysts start from a subjective definition of optimistic/pessimistic states or the scenarios are exogenously given to them. Portfolio performances over the different scenarios are then compared to explore the possible sensitivity of the financial assets to the shocks to the system. This study starts from the traditional sensitivity analysis where it considers

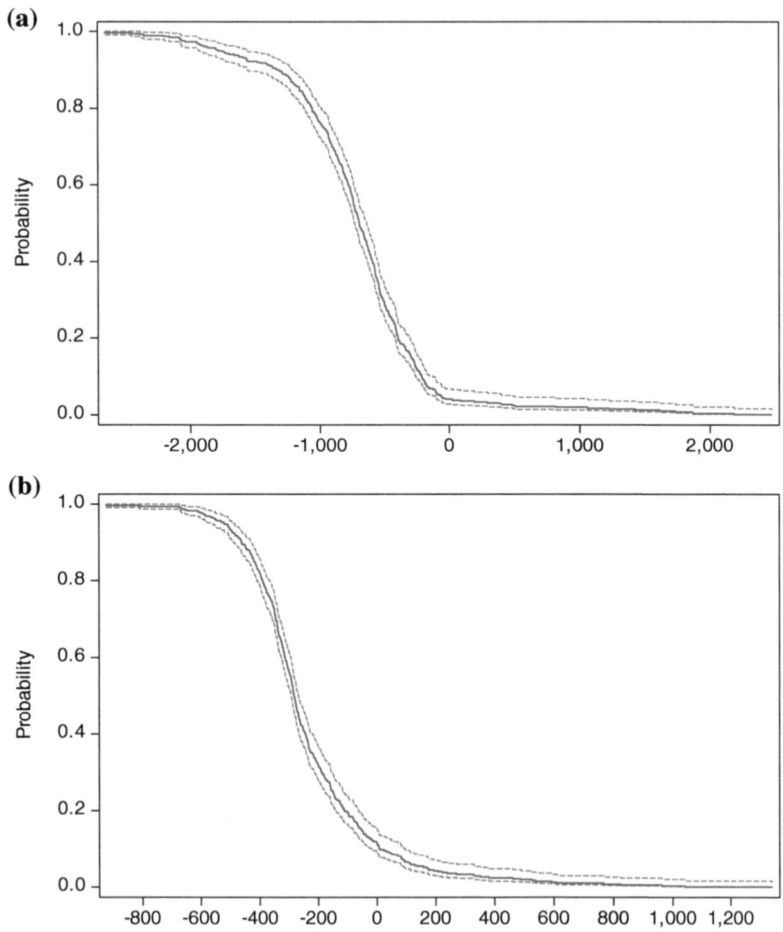

Fig. 2.13 Empirical survivor function for conditional correlation. **a** Green and market [prob (CCOR > 0) = 0.04]. **b** G75 and market [prob (CCOR > 0) = 0.11]. **c** G50 and market [prob (CCOR > 0) = 0.24]. **d** G25 and market [prob (CCOR > 0) = 0.71]. **e** Gray and market [prob (CCOR > 0) = 0.82]

financial crises to be exogenously given and defines crisis subjectively on the basis of movement in the benchmark index in Indian stock market. However, it extends itself to a survival analysis where researchers believe that stress depends on intrinsic vulnerability of a structure and it is a force exerted on by uncertainty and changing expectations of loss in financial markets. It is a continuous variable with a spectrum of values, where extreme values are called financial crises (Illing and Liu 2003). Hence, it concedes that it must be the performance of the asset itself, from where information should be extracted regarding the potential stresses that this asset might face, the asset's susceptibility to such stresses and its potential to survive or endure them.

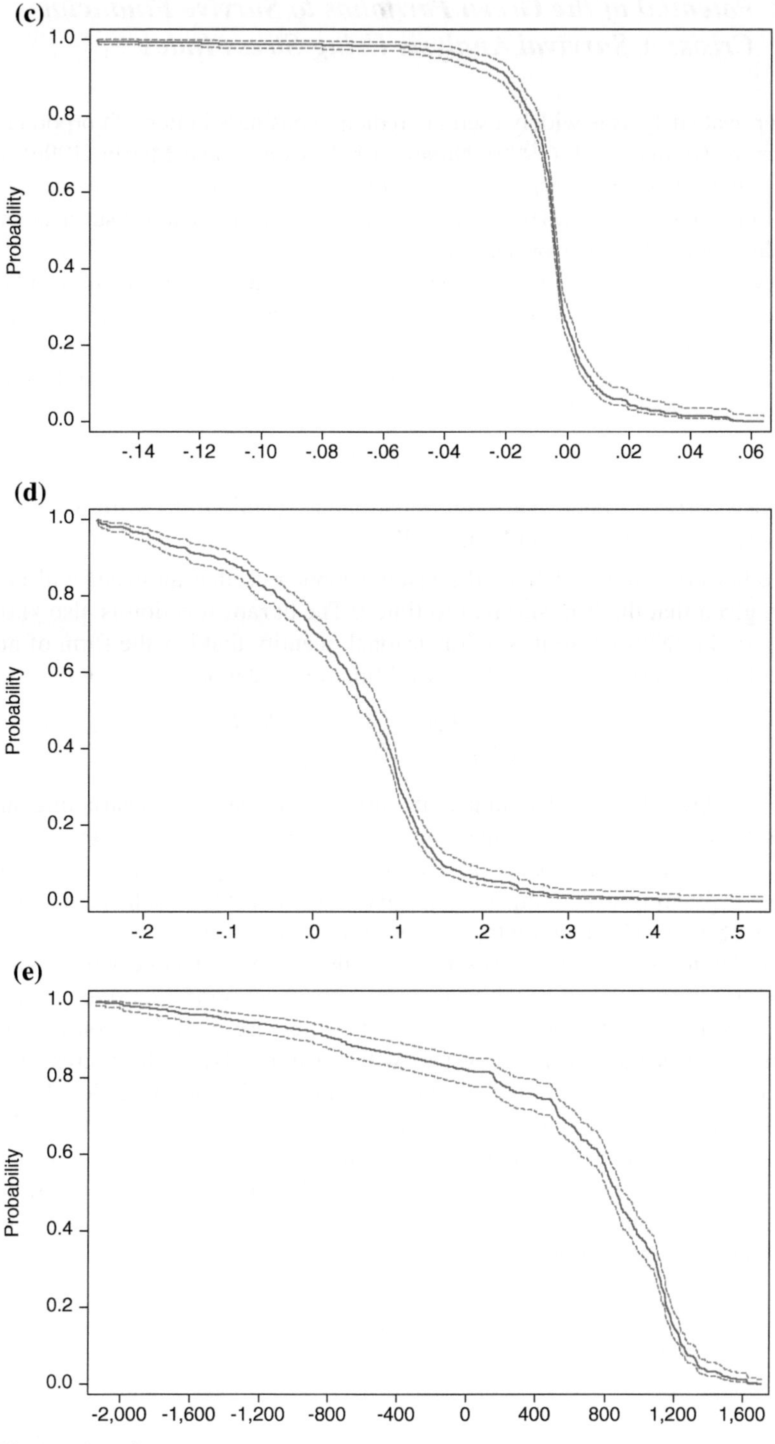

Fig. 2.13 (continued)

2.5.2 Potential of the Green Portfolios to Survive Financial Crisis: A Survival Analysis Using Stress Index

The survival analysis is widely used in predicting business failure of corporate firms. Studies by Thomas et al. (1999), Narain (1992), Cooper and Martin (1996), Lando (1997), and Jarrow and Turnbull (2000) are noteworthy in the context of predicting crisis in financial markets. However, studies are really rare that use survival analysis in evaluating portfolio performances.

Survival analysis involves two functions, namely the survivor function and hazard function. The survival function, $S(t)$, gives the probability that the time until an agent experiences the event, T, is greater than a given time t. Given that T is a random variable which defines the event time for some particular observation, then the survival function is defined as follows:

$$S(t) = \Pr(T > t) \tag{2.19}$$

Obviously, $S(t)$ is a decreasing right-continuous function of t with $S(0) = 1$ and $\lim_{t\to\infty} S(t) = 0$ (Kalbeisch and Prentice 2002).

The hazard function defines the instantaneous risk that an event will occur at time t, given that the firm survives to time t. The hazard function is also known as the "hazard rate" because it is a dimensional quantity that has the form of number of events per interval of time. The hazard function is defined as follows:

$$h(t) = \lim_{\Delta t\to 0} \frac{\Pr(t \le T < t + \Delta t | T \ge t)}{\Delta t} \tag{2.20}$$

There are three different techniques in survival analysis for constructing survival analysis models including nonparametric, semi-parametric, and parametric techniques. Nonparametric models are useful for preliminary analysis of survival data and for estimating and comparing survivor function. One such technique is the Kaplan–Meier (1958) method that has been used in this study.

In order to show how the portfolios could survive financial stress, the study constructs stress index for each of the green and gray portfolios. We concede that stress depends on intrinsic vulnerability of a structure as well as on the external factors. A shock pushed into a system will transform itself into a stress and then to a crisis if the system is intrinsically unstable or fragile. Traditional literature describes a financial system as fragile or intrinsically unstable by the weaknesses in financial conditions and/or in the structure of the financial system. The size of the shock and the interaction between financial system fragilities determine the level of stress. Illing and Liu (2003) considered the mechanism of shock transmission through a simple diagram.

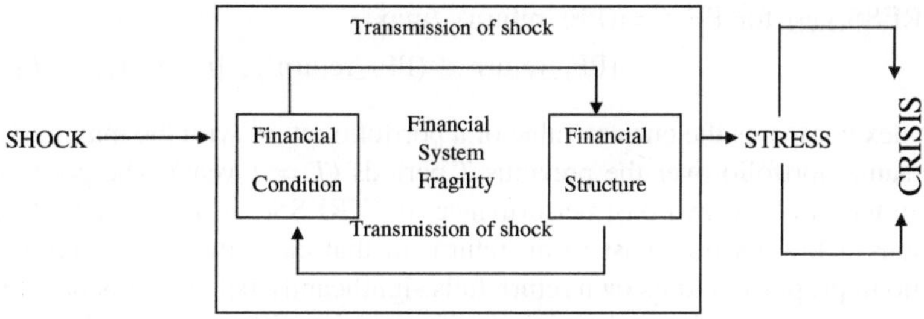

In literature, some stress indexes are available for the equity markets. This study follows the approach of Patel and Sarkar (1998) and of Vila (2000) with some modification to identify crises in the context of the constructed portfolios. Patel and Sarkar (1998) identified crisis in eight developed and fourteen emerging markets using the CMAX method which is a hybrid volatility loss measure. The CMAX method constructs the stress index as follows:

$$\text{CMAX} = X_t / \max\left[X \in \left(X_{t-j} | j = 0, 1, \ldots, T\right)\right]$$

where X_t is the stock index. The moving window is determined by T, and it is usually one to two years. Hence, CMAX compares the current value of a variable with its maximum value over the previous T periods. Vila (2000) used this method to identify periods of slide in the stock market. The trigger level is considered at either 1.5 or 2 standard deviations below the mean of the series.

This study defines stress in portfolio return in terms of the market as well as the portfolio itself. Hence, we consider two stress indexes: (i) $\text{STRESS}_{\text{MARKET}}$: an index which shows whether a portfolio is in stress in terms of the market and (ii) $\text{STRESS}_{\text{OWN}}$: an index which shows whether the portfolio is in stress in terms of its own performance. The two indexes are constructed as follows:

$\text{STRESS}_{\text{MARKET}}$: A portfolio is considered to be in crisis if it offers a return which is below 1.5 or 2 standard deviations of the past mean return in the market index. The stress index thus is same as a volatility loss measure and is constructed as follows:

$$\text{STRESS}_{\text{MARKET}} = (\text{PF\_return})_t / \max\left[\text{market\_return} \in (\text{market\_return})_{t-j} | j = 0, 1, \ldots, T\right] \quad (2.21)$$

where BSE SENSEX, the thirty-stock benchmark index of Indian economy, is taken as the proxy for the market.

The index compares the current value of a portfolio return with the maximum market return over the previous T periods ($T = 1$ year). The portfolio is in stress, if STRESS is less than 2 standard deviations below the mean of the market return. In that case, the current return of the portfolio in proportion to market return falls significantly below the historical market return.

$\text{STRESS}_{\text{OWN}}$: This index defines crisis as a situation when a portfolio offers a return which is 1.5 or 2 standard deviations below the past mean return of the same portfolio. The stress index thus is once again similar to a volatility loss measure and is constructed as follows:

$$\text{STRESS}_{\text{OWN}} \text{ for PF } 'i' = (\text{PF}_i\_\text{return})_t / \max [\text{PF}_i\_\text{return} \in (\text{PF}_i\_\text{return})_{t-j} | j = 0, 1, \ldots, T] \quad (2.22)$$

This index compares the current value of a portfolio return with the maximum return of the same portfolio over the previous T periods ($T = 1$ year). The portfolio is in stress in terms of its own past performance, if $\text{STRESS}_{\text{OWN}}$ is less than 2 standard deviations below its own past mean return. In that case, the current return of the portfolio in proportion to its own return falls significantly below its historical return.

The study now explores the periods of crisis for different portfolios using the STRESS indexes and tries to comment on the probability of each portfolio to survive the stress. The nature and the trends in the survival and hazard ratios for the green and gray portfolios are identical in terms of the two indexes. Hence, the study reports figures depicting survival probabilities and hazard ratios only once.

As is suggested by the figures, all the green portfolios have higher probability of surviving stress and lower hazard ratios compared to the gray one. Figure 2.14 shows the survival probabilities of different portfolios along with their trends, and Fig. 2.15 shows the hazard ratio lines for all the green and gray portfolios along with the respective trend lines of hazard. The gray portfolio has the maximum hazard ratio, followed by the G25, G50, and G75 portfolios. The trend hazard line for the 100 % green portfolio coincides with that of the G75 portfolio. On the other hand, the gray portfolio has the lowest probability of surviving financial stress, followed by G25, G50, and G75 portfolios. The probability of survival for the 100 % green portfolio once again coincides with that of the G75 portfolio. Thus, the green portfolios are intrinsically stronger than the gray portfolio in the sense that the greens could survive and endure stress in a better way. Moreover, the more green the portfolio, the stronger it is so far as surviving stress is concerned. There

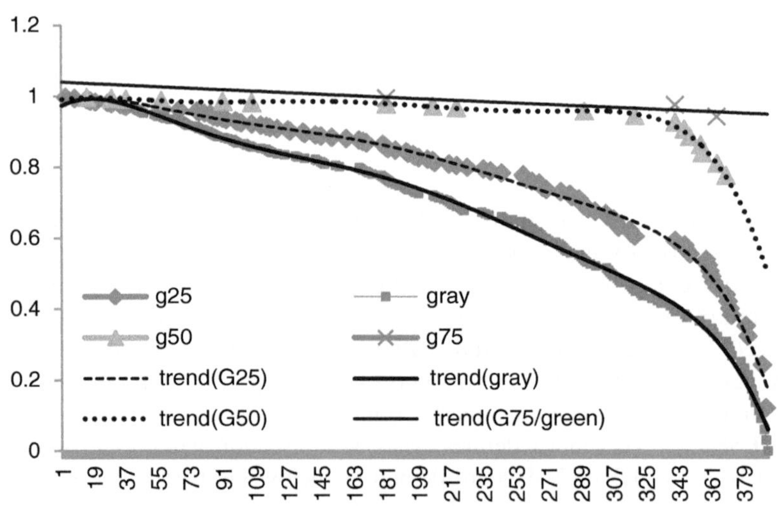

Fig. 2.14 Survival functions for the green and gray portfolios

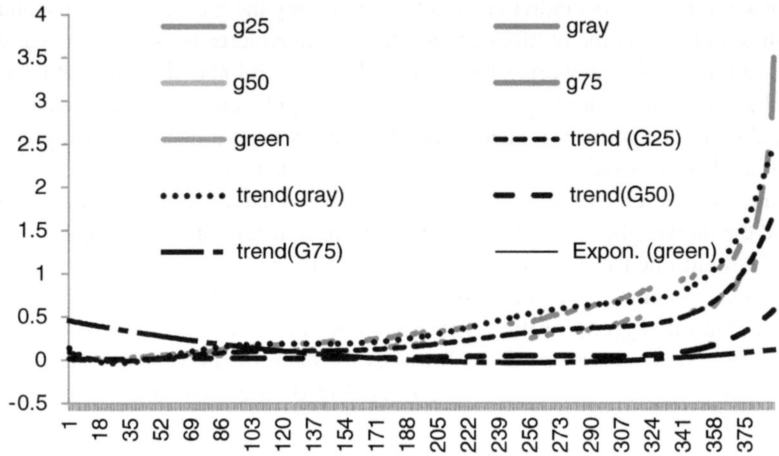

Fig. 2.15 Hazard ratios for the green and gray portfolios

is, however, one exception. The G75 portfolio, which is 75 % green, has the same survival probability and hazard ratio as the 100 % green portfolio. Thus, there might be a critical extent of "greenness," beyond which adding more green stocks in the portfolio does not improve the probability of surviving stress. This, however, does not mean that the 100 % green portfolio would be dominated by the G75. Investors still have an incentive to choose the global minimum variance "all-green" portfolio as it stochastically dominates G75 in terms of risk-adjusted return and has the lowest probability of offering positive conditional correlation with the market return.

While green portfolios could survive crisis originating from the market and from within itself, analysts may raise a pertinent issue, namely what affects the probability of surviving crisis? Is it mere sound company fundamentals or are there some other factors related to the intrinsic nature of the stocks constituting the portfolios? The study now delves deeper to explore factors that could yield protection against failure for the firms.

2.6 Factors Affecting Financial Stress: Greens Versus Grays

In the traditional literature, business failure prediction models make use of various statistical techniques in an attempt to estimate the bankruptcy probability of a firm using a set of covariates such as financial ratios and market-related variables (Beaver 1966; Altman 1968; Ohlson 1980; Zmijewski 1984; Whalen 1991; Laitinen and Luoma 1991; Shumway 2001; Chava and Jarrow 2004; Laitinen 2005; Jaggia and Thosar 2005; Gepp and Kumar 2008).

This section considers individual stocks constituting the green, semi-green, and gray portfolios, rather than the portfolios themselves, and explores possible factors affecting the probability of avoiding crisis for such stocks. The exploration is difficult to be considered at the portfolio level as it would be rather injudicious to define fundamental or financial ratios for portfolios. Hence, the study starts from a firm-level analysis where the probability of avoiding crisis for these firms is anticipated to depend on several covariates. While some of these covariates would be related to the company fundamentals and the market, the rest would reflect the intrinsic nature of the stock concerned.

As is mentioned earlier, the study selected twenty-five green and twenty-five gray stocks to construct green, semi-green, and gray portfolios. This section considers a combined group of these fifty-two stocks and constructs the stress indexes ($STRESS_{MARKET}$ as well as $STRESS_{OWN}$) for each of these on the basis of their past prices. The stress indexes are then used to define crisis and tranquil periods for each of these firms.

The dependent variable (*Y*) in the defined model then denotes the bankruptcy or otherwise of the firm. It is a dichotomous variable defined as follows:

$$Y = \begin{cases} 1 \text{ if the firm is in crisis} \\ 0 \text{ otherwise} \end{cases} \tag{2.23}$$

The study uses a Probit model that takes the form:

$$\Pr(Y = 1|X) = \phi(X'\beta) \tag{2.24}$$

Hence, the probability of being affected by crisis for a firm depends on a vector of regressors *X* that constituted of some factors that are assumed to influence the response variable *Y*. The parameters β are estimated by maximum likelihood. The traditional literature suggests few financial indicators that an investor will take into account while taking investment decisions. These financial ratios indicate a firm's financial health and its ability to fetch high return. The vector of regressors used in the study includes the following:

1. Variables related to company fundamentals:
   i. *Net profit margin* in percentage defined as profit after tax as a percentage of total operating revenue. This financial ratio is an indicator of *performance* of a firm;
   ii. *Dividend per share* defined as the dividends paid per unit of number of shares outstanding. This ratio indicates the *payout policy* of the firm and is relevant for the investors in calculating gains from investment;
   iii. Percentage *growth rate in earning per share* defined as the compound annual growth rate in EPS. This ratio measures the *actual growth* of the firm;
   iv. *Current ratio* defined as current asset in proportion to current liability. The ratio measures the *short-term solvency* or *liquidity* of the firm;
   v. *Debt equity ratio* defined as total debt in proportion to total equity indicates the *leverage* and hence the risk of the firm;
   vi. *Financial charge coverage ratio* is an indicator in the area of coverage and measures the firm's ability to meet financial obligations;

vii. *Asset turnover ratio* defined as total operating revenue generated per unit of asset of the firm indicates its *efficiency* to select and utilize assets;
viii. *Retention ratio* measures the proportion of net profit that is not distributed as dividends and reflects the firm's *perceived investment and growth opportunity* on the assumption that earnings retained will be invested in a profitable venture.

2. Variables related to market:
   i. *Average conditional correlation of a stock's return with the market return* is a variable related to market that measures the degree of time-varying association of a particular stock's return with the market movement. This measures the extent of market risk.
3. Variables related to the intrinsic nature of the stock:
   i. *Risk-adjusted return* measures the return that could be enjoyed over the risk-free rate after adjusting for the unique risk;
   ii. The extent of greenness of a stock is measured by a dichotomous variable that takes up the value 1 if the stock is green and is zero otherwise.

The study will consider two variations of the model. First, it would consider the factors affecting the probability of crisis using the $\text{STRESS}_{\text{OWN}}$ index to get an idea regarding the determinants of crisis where crisis is defined in terms of the stocks' own historical return. The same exercise will then be replicated using the $\text{STRESS}_{\text{MARKET}}$ index to gauge influence of different factors on the probability of crisis where crisis is defined in terms of the market's historical return. We summarize the main results in this section. The detailed results are depicted in Tables A.1 and A.2 in Appendix.

As is suggested by the LR statistic, the model is a good fit. When crisis is defined in terms of the stock's own historical return, the variables that could affect the probability of crisis are current ratio, net profit margin, retention ratio, and the dummy variable "green" that checks whether the stock in question is a green stock. The coefficients of each of these variables are significantly negative, implying a reduction in the probability of facing crisis with an increase in these variables. Out of these, the first three are related to the company fundamentals and its financial health. An increase in firm's short-term liquidity given by the increased current ratio will tend to reduce the probability of its stock to face crisis. The increased profitability of the firm and enhanced potential growth opportunity are also important factors in lowering the probability of crisis. More importantly, there is a single intrinsic nature of the stock, namely the extent of its greenness that is affecting the probability of crisis effectively. Increased greenness is lowering the probability of crisis. While the probability of crisis is being affected by short-term factors implying the myopic nature of the market, green stocks remain the obvious choice.

The results are modified when crisis is defined in terms of the historical returns in the market. The company fundamentals except for the current ratio cannot significantly affect the probability of crisis. The conditional correlation with the market and greenness, however, can significantly influence the probability of crisis. The probability of facing a crisis decreases significantly with an increase in liquidity, reduction in the degree of association with the market, and an increase in greenness of the particular stock in question.

Hence, stocks with high short-term liquidity, high profitability, and high perceived growth opportunity can avoid a crisis when it generates endogenously. However, company fundamentals have very limited role to play in the context of market risk. It is the liquidity and the lower degree of association with the market that could save the firms when crisis generates from within the market. The greens, however, remains the choice of the day. A green stock is always able to avoid crisis, be it generated exogenously or endogenously.

The study further uses the probability response curves to judge the effectiveness of greenness in influencing the change in the probability of avoiding crisis with change in the significant explanatory variables. The study considers how the probabilities of avoiding crisis for green and gray stocks differ for different values of retention ratio, net profit margin, and own stress index.

Figure 2.16 depicts the effect of greenness of stocks on the probability of suffering from crisis (when crisis generates endogenously) for various values of the stocks' net profitability margin. As the net profit margin or profitability of the firm increases, the probability of suffering from crisis for the green stocks falls significantly and gradually below the corresponding probabilities of the gray stocks. The gray stocks' probability of avoiding crisis, however, does not depend significantly on their profitability and remain at significantly higher level compared to that for the green stocks.

To establish further the greater impact of greenness on avoiding the probability of crisis, the study now considers the change in probability of crisis with respect to change in potential investment opportunity given by the retention ratio.

Figure 2.17 depicts the differential impact of greenness on the probability of crisis for different values of investment opportunity. Once again, the probability of suffering from crisis is more or less independent of changes in retention ratio for

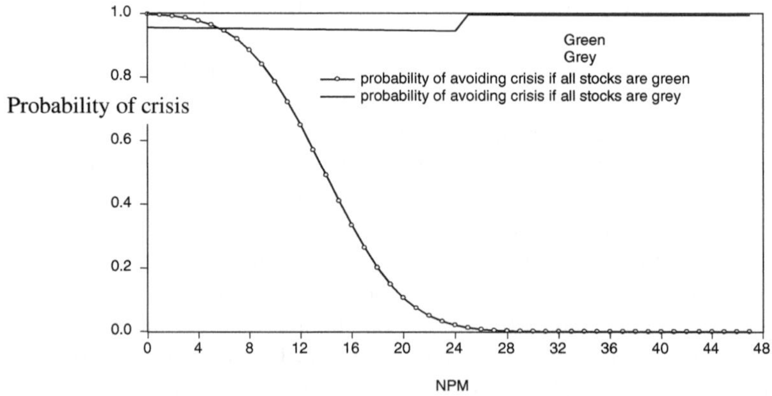

Fig. 2.16 Probability response curves for greens and grays with respect to NPM (crisis generates endogenously)

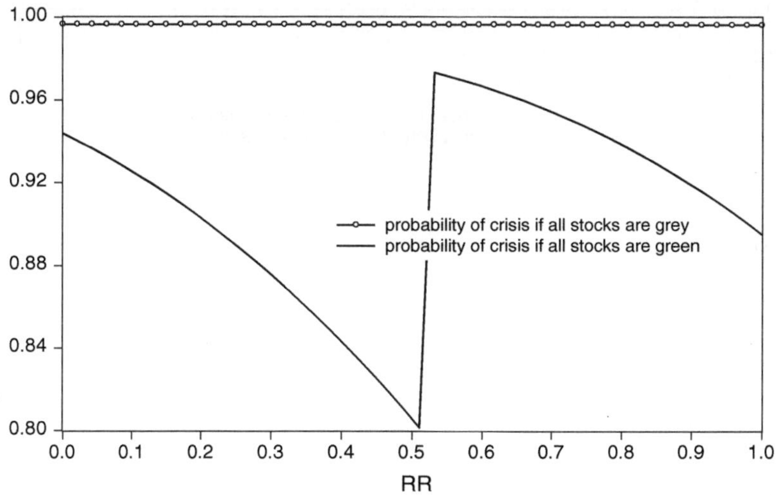

Fig. 2.17 Probability response curves for greens and grays with respect to retention ratio (crisis generates endogenously)

the gray stocks. The green stocks, however, have much lower probabilities of suffering crisis compared to their gray counterparts.

Similar results are obtained but with some exception when probability of suffering crisis for both types of stocks is analyzed with respect to changes in their own stress index. For lower values of stress index, the probability of crisis is significantly lower for the green stocks. However, there is a threshold level of shock beyond which the probabilities for the two groups merge (Fig. 2.18).

The probability response curves thus bring out significant impact of greenness on the probability of crisis for different values of fundamental variables. To summarize, green stocks have lower probability of suffering crisis for different levels of profitability and perceived growth opportunity. However, greenness matters only at the lower levels of endogenous shocks.

The exploration is now supplemented by an analysis of effectiveness of greenness in affecting the probability of surviving crisis generated from market. Once again, the probability response curves are drawn to explore the differential impact of greenness on the probability of crisis at different levels of significant explanatory variables, namely current ratio and conditional correlation with the market. The study is further extended to depict probability response curves with respect to different values of market stress index.

Figure 2.19 depicts differential impact of greenness of stocks on the probability of crisis at different levels of conditional correlation with the market. The probability of crisis for the gray stocks is significantly higher compared to that of the green stocks. The probabilities fluctuate as conditional correlation with the market changes. The green stocks' probability of suffering crisis is significantly lower. It declines with the increase in conditional correlation with the market and

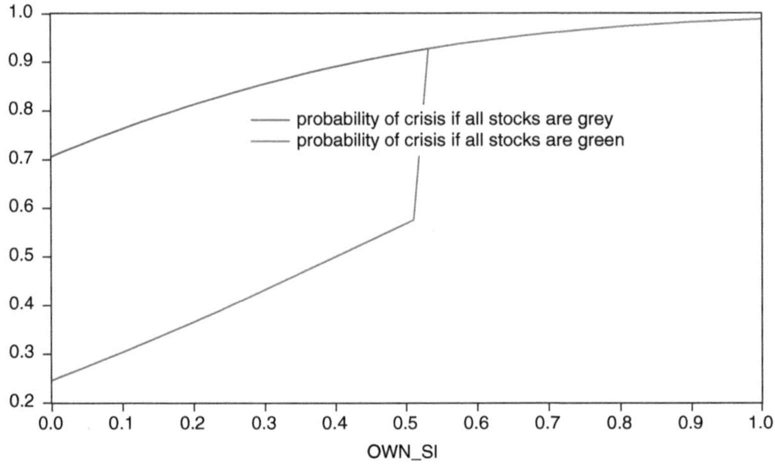

Fig. 2.18 Probability response curves for greens and grays with respect to own stress index

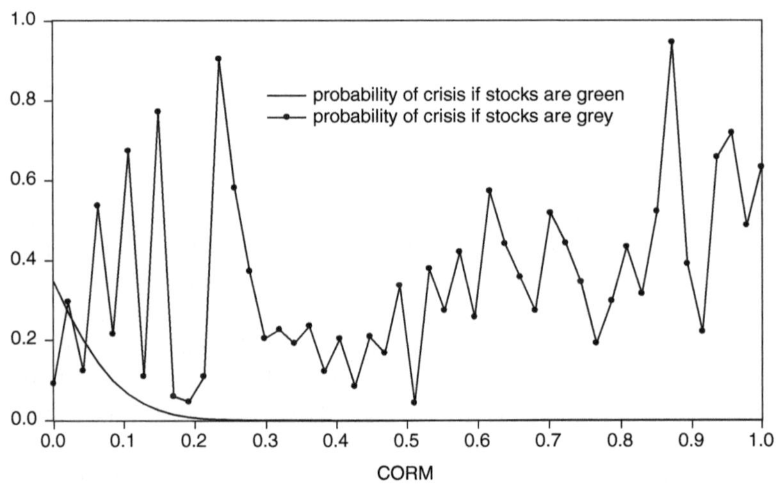

Fig. 2.19 Probability response curves for greens and grays with respect to conditional correlation with the market (crisis generates from the market)

eventually turns out to be zero. Hence, greenness significantly helps stocks to avoid market risk.

The study now considers the differential impact of greenness on the probability of crisis for different levels of short-term liquidity given by the current ratio. Once again, the probability of crisis for the stocks, if they are all gray, is significantly high and fluctuates along with the change in current ratio. The probability of crisis for all-green stocks, however, is more or less independent of the change in short-term liquidity (Figs. 2.20 and 2.21).

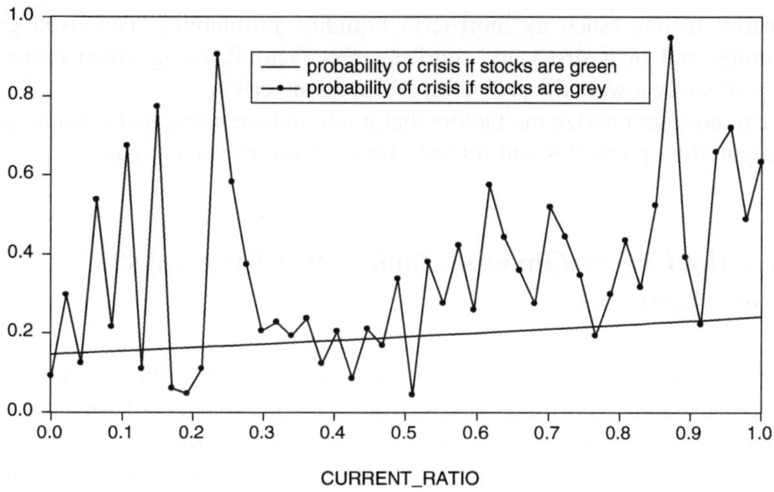

Fig. 2.20 Probability response curves for greens and grays with respect to current ratio (crisis generates from the market)

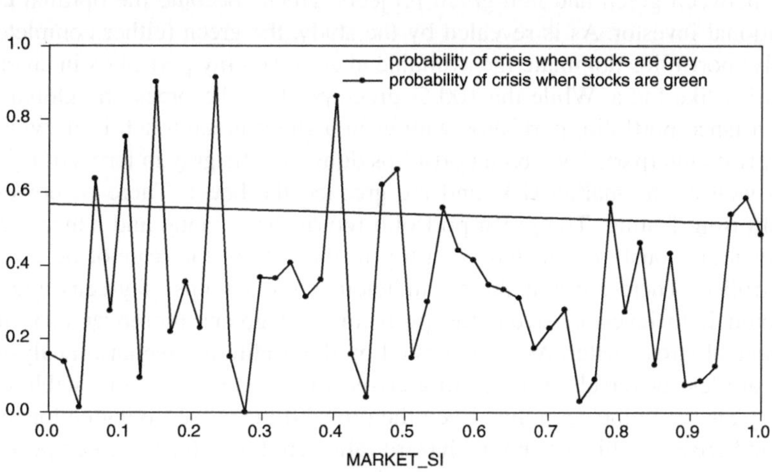

Fig. 2.21 Probability response curves for greens and grays with respect to market stress index (crisis generates from the market)

Slightly different results are obtained when we consider the effectiveness of being green in influencing the probability of crisis at different levels of market stress. The gray stocks have a probability of crisis which is high and more or less independent of the levels of market stress. The green stocks' probability of crisis fluctuates with market stress but is mostly lower than that of the gray portfolios.

The probability response curves thus bring out clearly the differential impact of being green and gray on the probability of avoiding crisis. For all levels of explanatory factors (such as short-term liquidity, profitability, perceived growth opportunity, and own stress and market stress factors), being green rather than gray significantly lowers the probability of avoiding crisis.

We can now summarize the factors that might induce investors to choose greens over gray in their portfolios and follow a less-carbon investment path.

2.7 Are the Greens Obvious Choice Over the Grays? Some Remarks

The present study is an exploration into an area where individual decision-making problems have significant implications from the point of view of society itself. It could hardly be denied that while following less-carbon investment path through increased investment in "green" projects are socially desirable, its implementation is not so easy. The task of the policy-makers, however, will be simplified if the imperative choice of the new "green" financial investment products is in fact obvious. This chapter delves specifically into this issue to examine whether given a choice between green and non-green projects, greens become the optimal choice of a rational investor. As is revealed by the study, the green (either completely or partially) portfolios dominate the available alternative gray portfolios in an emerging market like India. While the 100 % green portfolio becomes, the global minimum variance portfolio, portfolios with even a slight green touch in it, wins over their gray counterpart. The green portfolios dominate the gray in terms of the own-risk as well as the market risk, and the greener, the better. There is, however, a discomforting feature. The green portfolio returns are chaotic and hence are subject to endogenously originated volatility in the system that cannot be predicted or controlled. There are nonetheless judicious combinations of greens and grays that could help investors avoid the problem. The greens moreover can survive the financial stress better. As is suggested by the traditional scenario analysis, the greens are less adversely affected in a crisis situation and are more stable during recovery phases in the economy. Even the probabilities of surviving endogenously generated crises are higher and the hazard ratios are lower for the green portfolios. Although the study finds that there may be a critical balance between green and gray beyond which adding more green assets to a portfolio does not improve its probability of surviving financial stress, the "all-green" portfolio still dominates the others. More interestingly, it is not the company fundamentals but the extent of greenness that could effectively influence the probability of surviving crisis at the firm level. Hence, green is always preferred to gray, and more green is better than less green. Thus, following a less-carbon investment path is the most rational and obvious choice for the investors in the Indian market.

The study thus far brings about the justification for the risk-averse investors of choosing green portfolios over the grays. The techniques used in the analyses, however, do not explicitly bring in the consideration of the market. A financial analyst often concedes that while the fundamentals, the intrinsic stability, and the innate ability to avoid crises are some of the necessary criteria for choosing stocks, these are not sufficient. Investors, particularly, if they are myopic, will search for stocks that could offer returns that are significantly and consistently above the market return, or stated alternatively, that could "beat the market." While the speculators deem the ability to beat the market to be a sufficient criterion for choosing stocks, a rational investor will always look for a fundamentally strong stock that could beat the market. The study is now extended to introduce the market to judge the market performance of the otherwise "strong and viable" green stocks. Specifically, it considers the actual trading rule in the market to find out whether the fundamentally strong greens could actually offer a more profitable trading strategy, compared to their gray counterparts, which the investors could take advantage of. The existence of such momentum trading would establish the green's ability to beat the market consistently. This will, of course, put the efficient market hypothesis, a pillar on which traditional theories of finance rest, on trial. Nonetheless, the supremacy of greens over the grays will be avowed.

References

Akaike H (1974) A new look at the statistical model identification. IEEE Trans Autom Control 19(6):716–723

Altman EI (1968) Financial ratios, discriminant analysis, and the prediction of corporate bankruptcy. J Finan 23:589–609

Anderson G (1996) Nonparametric tests of stochastic dominance in income distributions. Econometrica 64:1183–1193

Bauwens L, S´ebastien L, Rombouts JVK (2006) Multivariate GARCH models: a survey. J Appl Econ 21:79–109

Baxter M, King RG (1999) Measuring business cycles: approximate band-pass filters for economic time series. Rev Econ Stat 81(4):575–593

Beaver WH (1966) Financial ratios as predictors of failure. empirical research in accounting: selected studies. J Account Res 4:71–127

Bollerslev T, Engle RF, Wooldridge JM (1988) A capital asset pricing model with time varying covariances. J Polit Econ 96:116–131

Brock W, Dechert W, Scheinkman J (1987) A test for independence based on the correlation dimension. Working paper, University of Wisconsin at Madison, University of Houston

Brown LD, Tony Cai T, DasGupta A (2001) Interval estimation for a binomial proportion. Stat Sci 16(2):101–133

Chakrabarti G, Sen C (2013) Momentum Trading on the Indian stock market. Springer, India

Chava S, Jarrow RA (2004) Bankruptcy prediction with industry effects. Rev Finan 8(4):537–569

Christian LJ, Fitzgerald TJ (2003) The band pass filter. Int Econ Rev 44(2):435–465

Cooper I, Martin M (1996) Default risk and derivative products. Appl Math Finan 353–374

Dardanoni V, Forcina A (1998) A unified approach to likelihood inference on stochastic orderings in a nonparametric context. J Am Stat Assoc 93:1112–1123

Dardanoni V, Forcina A (1999) Inference for Lorenz curve orderings. Econ J 2:49–75

Davidson R, Duclos Jean-Yves (2000) Statistical inference for stochastic dominance and for the measurement of poverty and inequality. Econometrica 68(6):1435–1464

de Goeij PC, Marquering W (2004) Modeling the conditional covariance between stock and bond returns: a multivariate GARCH approach. J Financ Econo 2(1):531–564

Eeckhoudt L, Schlesinger H, Tsetlin I (2008) Appointing of risk via stochastic Dominance. Working paper

Fraser AM, Swinney HL (1986) Independent coordinates for strange attractors from mutual information. Phys Rev A 33:1134–1140

Gepp A, Kumar K (2008) The role of survival analysis in financial distress prediction. Int Res J Finan Econ 16:1450–2887

Giorgi De E (2002) Reward-risk portfolio selection and stochastic dominance. Working paper

Giorgi ED, Post T (2004) Second order stochastic dominance, reward-risk portfolio selection and the CAPM. Working paper

Gloy BA, Baker TG (1999) Evaluating risk management strategies using stochastic dominance with a risk free asset. AAEA Annual Meetings, Nashville, Tennessee, USA

Hodder JE, Jackwerth JC, Kolokolova O (2009) Improved portfolio choice using second order stochastic dominance. Working paper

Hyndman RJ, Fan Y (1996) Sample quantiles in statistical packages. Am Stat 50(4):361–365

Illing M, Liu Y (2003) An index of financial stress for Canada. Working Paper 2003-14, Bank of Canada

Ingersoll JE (1987) Theory of financial decision making. Rowman & Littlefield, Maryland

Jaggia S, Thosar S (2005) Survival analysis with artificially constructed events. Rev Account Finan 4(4):34–49

Jarrow RA, Turnbull SM (2000) The intersection of market and credit risk. J Banking Finan 24:271–299

Kahneman D, Tversky A (1979) Prospect theory: an analysis of decisions under risk. Econometrica 47:263–291

Kalbeisch JD, Prentice RL (2002) The statistical analysis of failure time data. Wiley, New York

Kantz H, Schriber T (2004) Nonlinear time series analysis, 2nd edn. Cambridge University Press, Cambridge

Kaplan DT, Glass L (1992) Direct test for determinism in a time series. Phys Rev Lett 68:427–430

Kaplan E, Meier P (1958) Nonparametric Estimation from Incomplete Observations. J Am Stat Assoc 53:457–481

Karunanayake I, Valadkhani A, O'Brien M (2009) Modelling Australian stock market volatility: a multivariate GARCH approach. Working Paper 09-11, Department of Economics, University of Wollongong, 15 p

Kaur A, Rao BLSP, Singh H (1994) Testing for second order stochastic dominance of two distributions. Econ Theor 10:849–866

Kennel MB, Brown R, Abarbanel HD (1992) Determining embedding dimension for phase space reconstruction using a geometrical construction. Phys Rev A 45:3403–3411

Klecan L, McFadden R, McFadden D (1991) A robust test for stochastic dominance. Working Paper, MIT and Cornerstone Research

Kodba S, Perc M, Marhl M (2004) Detecting chaos from a time series. Eur J Phys 26:205–215

Kuosmanen T (2001) Stochastic dominance efficiency tests under diversification. Working paper

Kuosmanen T (2004) Efficient diversification according to stochastic dominance criteria. Manage Sci 50:1390–1406

Laitinen EK (2005) Survival analysis and financial distress prediction: finish evidence. Rev Account Finan 4(4):76–90

Laitinen EK, Luoma M (1991) Survival analysis as a tool for company failure prediction. Int J Manage Sci 19(6):673–678

Lando D (1997) Modelling bonds and derivatives with credit risk. In: Dempster M, Pliska S (eds) Mathematics of financial derivatives. Cambridge University Press, Cambridge, pp 369–393

Levy H (1992) Stochastic dominance and expected utility: survey and analysis. Manage Sci 38:555–593

Levy H (2006) Stochastic dominance: Investment decision making under uncertainty, 2nd edn. Springer Science and Business Media, Inc., New York

Markowitz H (1952) Portfolio selection. J Finan 7:77–91

McFadden D (1989) Testing for stochastic dominance. In: Fomby TB, Seo TK (eds) Studies in the economics of uncertainty. Springer, New York, pp 113–132

Meyer TO, Xiaoming Li, Lawrence CR (2004) Comparing mean variance tests with stochastic dominance when assessing international portfolio diversification benefits. Working paper

Milos K, Chovakec P (2008) A second-order stochastic dominance portfolio efficiency measure. Kybernetika 44:243–258

Narain B (1992) Survival analysis and the credit granting decision. In: Thomas LC, Crook JN, Edelman DB (eds) Credit scoring and credit control. Oxford University Press, Oxford, pp 109–121

Nelson DB (1991) Conditional heteroskedasticity in asset returns: a new approach. Econometrica 59(2):347–370

Ng P, Wong WK, Xiao Z (2011) Stochastic Dominance via quantile regression. Working Paper series, 11-01, Northern Arizona University

Ohlson JA (1980) Financial ratios and the probabilistic prediction of bankruptcy. J Account Res 18(1):109–131

Patel S, Sarkar A (1998) Crises in developed and emerging stock markets. Finan Anal J (Nov/ Dec):50–61

Perc M (2005) The dynamics of human gait. Eur J Phys 26:525–534

Post T (2003) Empirical tests for stochastic dominance efficiency. J Finan 58:1905–1931

Post GV, Diltz JD (1986) A stochastic dominance approach to risk analysis of computer systems. MIS Quart 10:363–375

Rhode C, Morari M (1997) False-nearest-neighbors algorithm and noise-corrupted time-series. Phys Rev E 55(5):6162–6170

Rothschild M, Stiglitz J (1970) Increasing risk I: a definition. J Econ Theor 2:225–243

Sarkar A, Chakrabarti G, Sen C (2013) Volatility, long memory and chaos: a discussion on some "stylized facts" in financial markets with a focus on high frequency data. Springer, India

Shumway T (2001) Forecasting Bankruptcy more accurately: a simple hazard model. J Bus 74(1):101–124

Sriboonchitta S, Wong WK, Dhompongsa S, Nguyen HT (2010) Stochastic dominance and applications to finance, risk and economics. Taylor and Francis Group, Boca Raton

Takens, F (1981) Detecting strange attractors in turbulence. Lecture notes in mathematics, pp 366–381

Thomas, LC, Banasik J, Crook JN (1999) Not if but when loans default. J Oper Res Soc 50:1185–1190

Vila A (2000) Asset price crises and banking crises: some empirical evidence. BIS Conf Papers 8(3):232–252

Whalen GA (1991) Proportional hazards model of bank failure: an examination of its usefulness as an early warning tool. Federal Reserve Bank of Cleveland. Econ Rev Q 1:21–31

Wilson EB (1927) Probable inference, the law of succession, and statistical inference. J Am Stat Assoc, 209–221

Wirch JL, Hardy MR (2001) Distortion risk measures: coherence and stochastic dominance. Working paper

Wolf A, Swift JB, Swinney HL, Vastano JA (1985) Determining Lyapunov exponents from a time series. Physica D 16:285–317

Zmijewski ME (1984) Methodological issues related to the estimation of financial distress prediction models. J Account Res 22:59–82

References

Chapter 3
Profits Are Forever: A Green Momentum Strategy Perspective

Abstract This chapter attempts to examine the possibility of receiving consistently above-average returns for two "pure" (100 % green, 100 % gray) and three "hybrid" (25 % green, 50 % green, and 75 % green) portfolios. A popular tool of investment decision making has been momentum trading strategies. This chapter makes use of suitable momentum trading strategy to examine the investment-worthiness of the green, part green, and gray portfolios. The empirical analysis starts with an examination of long-term memory in the portfolio returns. Both graphical as well as statistical results suggest that only 100 % green and 100 % gray portfolios exhibit significant long-term memory. The study delves deeper and investigates the presence of any possible trading strategy in the portfolio returns. As the result suggests, only 100 % green and 100 % gray portfolio returns are characterized by a long run moving average-based trading strategy. Most importantly, the 100 % green portfolio leads to a higher return than the 100 % gray portfolio, reinforcing the investment-worthiness of green assets.

Keywords Momentum trading strategy · Long memory · R/S statistics · Moving average · Optimum trading rule

3.1 Beating the Market—End of a Myth?

Trading strategies in the stock market is essentially as old as the stock market itself (Charoenwong 2012). The term "trading strategy" refers to investment strategies with an aim of earning consistently high return from the market by attempting to forecast the price of a financial instrument based on its historical prices. Technical analysis lends itself to investment enthusiasts and researchers alike because of the predictability in the stock returns, which is stark in contrast to the existing notion of an efficient market. Presence of trading rule ensures that an above-average return is consistently possible or in other words, the market can be beaten consistently. An efficient market exhibits random walk, and if the market is characterized by random walk, trading strategies will be rendered useless. Brock et al. (1992) was first to

G. Chakrabarti and C. Sen, *Green Investing*, SpringerBriefs in Finance,
DOI 10.1007/978-81-322-2026-8_3

provide evidence of predictability of stock prices based on the past moving averages. They tasted two simple trading strategies, moving average, and trading range break, in the Dow Jones for 100 years of data between 1897 and 1986. Their study exhibited significant trace of predictive power of the moving average strategy. In more recent years, Lo et al. (2000) applied a technical pattern recognition model using nonparametric kernel regression to US stocks for 34 years. Their results suggest that several technical indicators are indeed proved to be profitable in nature. Neely et al. (2013) used technical indicators to verify their predictive power. The results reveal that the indicators have "significant in-sample and out-of-sample forecasting power." In recent past, there has been a whole body of literature on possible momentum strategies in the market. Jegadeesh (1990), Lo and MacKinlay (1988), and Fama and French (1988) showed that the market is characterized by negative correlation among stock returns in the long run but positive correlation in the short run, which implies that there is indeed a scope of profitable momentum strategies in the short run. Existence of positive autocorrelation among stocks in the short run and hence relevance of momentum strategies is also documented by Jegadeesh and Titman (1993, 1995). Lee and Swaminathan (2000) documented that momentum strategies are successful in the US stock market. Rouwenhorst (1998) considered 20 emerging stock markets and found significantly positive results when momentum strategies are implemented in all countries. Hameed and Yuanto (2002) showed existence of profitable momentum strategies in some selected Asian stock markets. Success of momentum strategies are also documented in the studies of Conrad and Kaul (1998), Richards (1997) and Liu et al. (1999).

3.2 Technical Trading Rules: A Review of the Alternative Methodologies

There are a number of technical trading strategies, ranging from relatively simple to very complex. This section discusses some of the strategies available to investors. The best possible trading strategy is often a matter of debate. However, according to Mashaushi (2006), "The choice of trading rules is a subject related to data snooping and spurious results. To avoid these problems we choose, (1) trading rules that are most widely used in the industry and (2) those that are simple to implement". Thus, study follows the above rule.

3.2.1 *Filter Strategies*

The simplest of trading strategies, this trading rule says that an investor should buy and hold a stock if its price increases x % above a previous low and sell the stock if the price falls x % below the previous high (Alexander 1961). The filter rule can be expressed as the following:

$$\left.\begin{array}{l}\text{If}\left(1-\dfrac{P_t}{P_{t-1}}\right) > \dfrac{x}{100}\text{then buy/hold}\\ \text{If}\left(1-\dfrac{P_t}{P_{t-1}}\right) < \dfrac{x}{100}\text{then sell/stay out}\end{array}\right\} \tag{3.1}$$

This strategy stems from the assumption that prices are serially correlated and characterized by long-term memory, which lends itself to the fact that a high stock price is likely to be followed by a higher than average stock price.[1]

Studies have considered a range of values for x, from 0.5 to 50 % in an attempt to find a strategy that returns consistently high profits (Brock et al. 1992; Bessembinder and Chan 1995). However, several studies have found evidence that although a suitable filter rule provides profit over and above the general buy and sell strategy, when accounted for trading costs, the simple buy and sell strategy actually returns higher profit than a filter strategy (Hudson et al. 1996; Ready 1997; Fama and Blume 1966). However, Harvey (1955) suggests that a trading rule is more likely to work in an emerging capital market due to its predictable nature.

3.2.2 Moving Average Rule

The moving average trading rule also is based on the assumption of positive serial correlation in stock prices. An increasing moving average with prices above the moving average is considered to be an upward trend. If the moving average is decreasing and the prices are below the moving average, it is considered to be a downward trend (Person 2007).

$$\left.\begin{array}{l}\text{If } \mathrm{MA}_t^s > \mathrm{MA}_t^l \text{ then buy/hold}\\ \text{If } \mathrm{MA}_t^s < \mathrm{MA}_t^l \text{ the sell/stay out}\end{array}\right\} \tag{3.2}$$

where

$$\mathrm{MA}_t^s = \frac{1}{k}\sum_{i=t-k}^{t} P_i$$

When the short moving average moves above the long moving average, an investor should buy and hold the stock till the moving averages are equal. To find the optimum trading rule, moving averages with varying lengths should be considered. Up to 2 weeks are considered to be short-term average, 14–30 days are considered to be medium term average and above 30 days are considered to be long-term averages.

[1] *Source* http://www.finance-trading-times.com/2007/12/trading-strategies-short-term-filter.html.

This research uses the moving average method as the trading strategy of choice. A detailed discussion of the methodology is provided in the next section.

3.2.3 Momentum Strategy Based on a Simple Regression Model

A momentum strategy based on a simple regression model helps understand the "price continuation" according to Moskowitz et al. (2012). In order to construct the regression model, the excess risk adjusted return r_t^s for instrument s in month t is regressed on its own h period lagged values, appropriately scaled by respective past periods volatility.

The regression equation can be written as:

$$\frac{r_t^s}{\sigma_{t-1}^s} = \alpha + \beta_h r_{t-h}^s / \sigma_{t-h-1}^s + \varepsilon_t^s \qquad (3.3)$$

A significantly positive β signifies positive impact of past risk adjusted returns on current period's risk adjusted return. In other words, a positive value of the t-statistic would be enough to ensure that past returns can successfully predict current returns. A negative t-value, on the other hand, would ensure value reversal.

3.2.4 Cross-sectional Momentum Strategy (XSMOM)

The cross-sectional momentum strategy (Jegadeesh and Titman 1993; Lewellen 2002; Asness et al. 2009) is based on the performance of the instruments in a decile portfolios constructed according to each instruments past performance. The investor should go long on the top 10 % instruments and short on the bottom 10 %.

The steps to construct a cross-sectional momentum strategy according to Moskowitz et al. (2012) is provided here step by step:

Step 1. The portfolio weight of a particular instrument i in period t is

$$w_t^i = \frac{1}{N}(r_{t-1;t}^i - r_{t-1;t}^{\mathrm{EW}})$$

where $r_{t-1:t}^i$ is the return of the instrument i and $r_{t-1:t}^{\mathrm{EW}}$ is the return on the equally weighted index, $r_{t-1:t}^{\mathrm{EW}}$

Step 2. The portfoliov return can be expressed as

$$r_{t,t+1}^{\mathrm{XS}} = \frac{1}{N}\sum_{i=1}^{N} w_t^{\mathrm{XS},i} r_{t,t+1}^i$$

Step 3. The expected return of the cross-sectional momentum can be decomposed as:

$$E\left[\pi^{XS}_{t:t+1}\right] = \frac{\mathrm{tr}(\Omega)}{N} - \frac{\iota'\Omega\iota}{N^2} + \sigma^2_\mu$$
$$= \frac{N-1}{N^2}\mathrm{tr}(\Omega) - \frac{1}{N^2}\left[\iota'\Omega\iota - \mathrm{tr}(\Omega)\right] + \sigma^2_\mu \tag{3.4}$$

where tr is the trace of the matrix,

1	is a $(N \times 1)$ vector of 1,
$\iota'\Omega\iota/N^2$	is the autocovariance
μ	is the expected return,
Ω	is the autocovariance matrices
σ^2_μ	is the cross-sectional variance of the return.

The Eq. (3.4) implies that there are three sources of cross-sectional momentum profit. A positive autocovariance (first part) implies that a positive return in the lookback period will lead to a subsequent positive return in the holding period. The second term refers to a negative cross-sectional variance, i.e., the likelihood of a security's earning a more than expected return will be ensured by the likelihood of another security's doing less than expected. The last term suggests that the cross-sectional variance of the return is also a source of momentum profit (Johansen and Villaddsen 2013).

3.2.5 Time Series Momentum Strategy (TSMOM)

TSMOM strategies can be constructed in a number of ways by changing the lookback period and holding period. The lookback period (*L*) refers to the time period an investor takes into consideration before forming a portfolio and holding period (*H*) is the time the investor holds the portfolio before going short. Therefore, an alternate regression equation can be formulated, with just the sign of the previous periods' returns.

$$\frac{r^s_t}{\sigma^s_{t-1}} = \alpha + \beta_h \mathrm{sign}(r^s_{t-h}) + \varepsilon^s_t \tag{3.5}$$

The return is scaled by the previous period's volatility in order to make it volatility independent. A positive return in the previous period will ensure a "+" sign, while a negative return will result in a "−" sign and a reversal.

The momentum strategy as formulated by Moskowitz et al. (2012) is that for each instrument, the investor should go long if the return in the lookback period is positive and should go short if the return in the lookback period is negative.

The TSMOM factor for an instrument *s* at time *t* is, therefore:

$$r^{TS,s}_{t:t+H} = \mathrm{sign}(r^s_{t-L:t})\frac{40\,\%}{\sigma^s_t} r^s_{t:t+H} \tag{3.6}$$

where 40 % is used as a scaling factor.

The aggregate return of the strategy that diversifies across all the s_t instruments that are available at time t is

$$\pi_{t:t+H}^{\mathrm{TS},s} = \frac{1}{S_t}\sum_{s=1}^{s_t} \operatorname{sign}(r_{t-L:t}^{s})\frac{40\,\%}{\sigma_t^s} r_{t:t+H}^{s} \tag{3.7}$$

where π is the portfolio return over the holding period.

Moskowitz et al. (2012) used a regression model to estimate the "abnormal performance evaluation" of the trading strategies.

$$\begin{aligned}\pi_{t:t+H}^{\mathrm{TS},s} &= \alpha + \beta_1 \mathrm{MKT}_t + \beta_2 \mathrm{BOND}_t + \beta_3 \mathrm{GSCI}_t + s\mathrm{SMB}_t \\ &\quad + h\mathrm{HML}_t + h\mathrm{UMD}_t + \varepsilon_t\end{aligned} \tag{3.8}$$

where MKT_treturn of the MSCI World Index

BOND_t return of the Barclays Aggregate Bond Index
GSCI_t return of the S&P Goldman Sachs Commodity Index
SMB_t mimicking returns for the size factor as suggested by Fama and French (1993)
HML_t mimicking returns for the value factor as suggested by Fama and French (1993)
UMD_t one year cross-sectional momentum factor (Carhart 1997)

Similar to the cross-sectional momentum, three TSMOM can also be decomposed. If the portfolio weight w_i for an asset i in month t is defined as:

$$w_t^i = \frac{1}{N} r_{t-1;t}^i$$

The expected return can be expressed as:

$$E\left[\pi_{t:t+1}^{\mathrm{TS}}\right] = \frac{tr(\Omega)}{N} + \frac{\mu^{'}\mu}{N} \tag{3.9}$$

The above equation suggests that the expected return of a portfolio can be decomposed into two parts, the first parts is the average autocovariance of the instruments and the second part is the average squared mean excess return.

3.3 Optimal Trading Rules

In the next section, the study continues with identifying possible moving average trading rules. And for this purpose, the moving average rule has been selected. According to the trading rule, an investor should go long when price is above some moving average of historical prices and go short when price falls below some moving average. This section attempts to find out an optimum trading rule for each portfolio constructed (Green, 25 % green, 50 % green, 75 % green and gray).

This will give an insight into how the market can be "beaten" or a higher than expected return can be predicted. The ramification is, manifold.

The analysis will be done in three stages. The details are given below.

3.3.1 Detailed Methodology

Step 1. The variable underlying would be the daily risk adjusted return series for each portfolio.

Step 2. Suitable short run, medium run, and long run moving averages are to be constructed.

Step 3. A general regression model will be constructed which will work as a benchmark for a profitability comparison with trading rule.

Step 4. The study considers 3 and 7 days as short run, 14, 21, and 30 days as medium run, and 60, 100, 150, 200, and 270 days as long run, respectively. All possible combinations of short run, medium run, and long run signals are constructed such as buy37(short run combinations), buy314, buy714, buy321, buy721, buy330, buy730 (short run–medium run combination), buy1421, buy1430, buy2130 (medium run–medium run combination), buy360, buy3100, buy3150, buy3200, buy3270, buy760, buy7100, buy7150, buy7200, buy7270 (short run–long run combination), buy1460, buy14100, buy14150, buy14200, buy14270, buy2160, buy21100, buy21200, buy21270, buy3060, buy30100, buy30150, buy30200, buy30270 (medium run–long run combination), and buy60100, buy60150, buy60200, buy60270, buy100150, buy100200, buy100270, buy150200, buy150270, buy200270 (long run–long run combination). In reality, an infinite number of combinations can be generated but this study considers till 270 days which, in a high frequency market such as stock market, is significantly long run.

All the buying signals are dummy variable taking a value of 1 if the price is above the *n*th day moving average of historical prices and zero otherwise.

Step 5. Regress return on a constant and a lagged value of buy signal. The estimated slope coefficient will give the possible profit. A significantly positive slope coefficient, which is higher than the intercept as well as the coefficient of the general buy–sell regression hints toward existence of a profitable trading rule,

Step 6. All possible buying signals (the set under consideration) are taken into account and the signal with highest profitability shows the optimum trading strategy.

There are, however, some limitations of this study. For example, (i) the bid—ask spread is not considered, as the data is not available, (ii) intraday or tick-by-tick price, rather than daily price would have given a more accurate result, and (iii) possible idiosyncrasies may exist, as India is still an emerging market.

3.3.2 Finding the Optimum Trading Rule

Momentum, or in other words, positive serial correlation is a common phenomenon in financial asset returns. As the main objective of a momentum trader is to make profit, it is very important from a traders perspective to understand when to ride the momentum train and when to get off it. Conceptually, a successful momentum strategy depends mainly on the ability of past prices to predict the future prices. Therefore, significantly high degree of autocorrelation can be expected in the series. The next section takes a look at the autocorrelation.

3.3.2.1 Visual Analysis of Autocorrelation

An inefficient market is often characterized by long-term memory and traders usually exploit this characteristic to their benefit. Before we move into analyzing the market trend and trading rules, it is beneficial to examine whether the underlying series is characterized by long-term memory, or, in other words, whether a shock propagated into the system remains within it for a long time or dies down very quickly. One simple way to test it is checking for autocorrelations. Using the correllogram function, the autocorrelations for each of the twelve return series are generated for 36 lags. Return is defined at $R_t = \ln(P_t/P_{t-1})$ where P_t is the price on *t*th day. Interestingly, even after 36 lags, the autocorrelations do not die down. They still remain significantly greater than zero. The autocorrelation functions are plotted in Figs. 3.1, 3.2, 3.3, 3.4, and 3.5.

Based on the visual evidence above, it can be said that only the two "pure" portfolio returns (green and gray) are characterized by slowly decaying autocorrelation. Even after 36 lags, the autocorrelation does not fall to zero for these two portfolio returns. However, there is no discernible evidence for decaying autocorrelation for the three "hybrid" portfolios. In order to make a conclusive remark, it is essential to understand whether the hybrid portfolio returns are indeed not influenced by past returns.

To have more concrete evidence regarding the long run dependence, the five series are subjected through tests for long-term memory.

3.3.2.2 Long Memory in the Return Series: The R/S Statistics and Modified R/S Statistics

Long memory is said to exist in the market if there is significant autocorrelation between widely separated observations. It is a process in which the autocorrelation function decays asymptotically as a function of the time lag. As a result of the presence of long-term memory, present outcomes can be significantly affected by the outcomes from distant past. Long memory has long made its presence felt in the areas of natural sciences. So much so, it owes its genesis to a natural scientist, Harold Edwin Hurst. He had first designed a test for long-term dependence

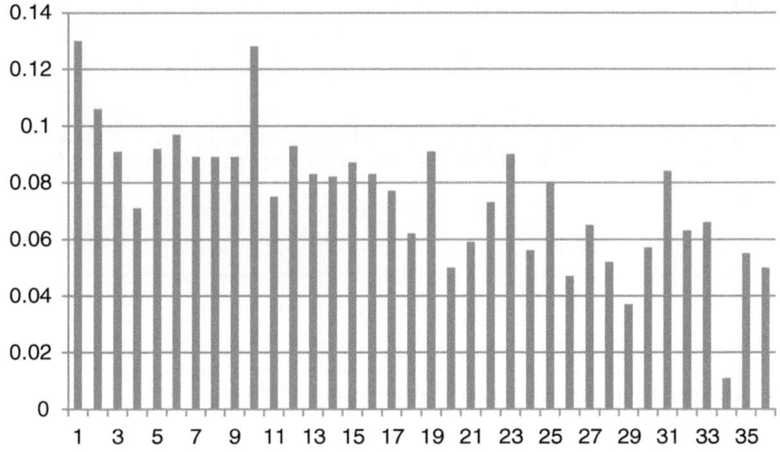

Fig. 3.1 Autocorrelation function for green portfolio (36 lags)

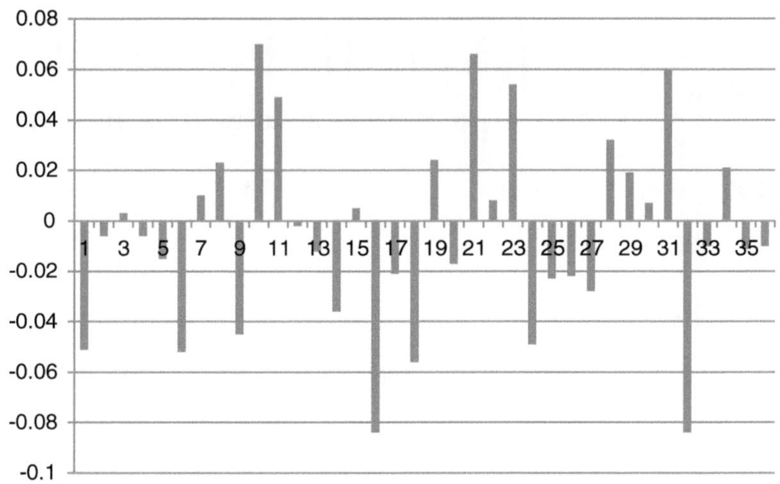

Fig. 3.2 Autocorrelation function for 25 % green portfolio (36 lags)

to predict the pattern of flooding by the river Nile known as the *rescaled range statistic* or the *R/S statistic*. This statistics is known as Rescaled Range statistic or simply, R/S statistic (Hurst 1951).

The study continues to look for more evidences for long memory, and in this pursuit, it employs a more traditional and widely used test called the Rescaled Range Statistics or the R/S Statistics. This test was first used by H.E. Hurst in 1951 to test long memory in the pattern of flooding of river Nile. It was later modified by Mandelbrot (1972, 1975).

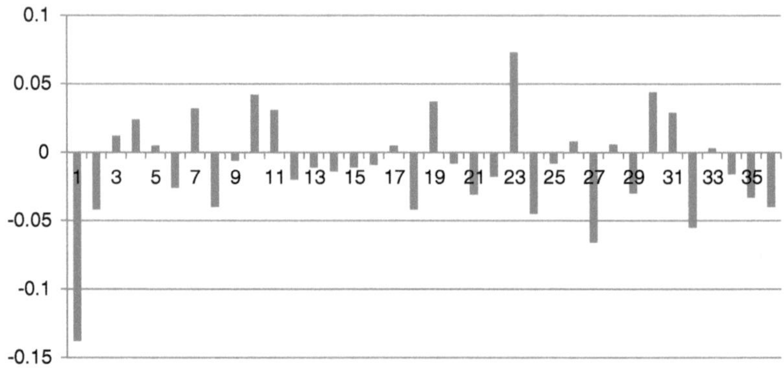

Fig. 3.3 Autocorrelation function for 50 % green portfolio (36 lags)

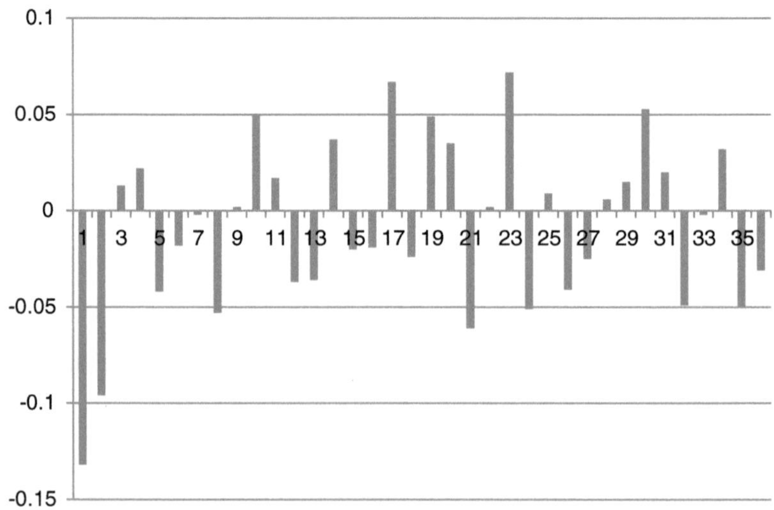

Fig. 3.4 Autocorrelation function for 75 % green portfolio (36 lags)

The R/S statistic is defined as:

$$Q_T = \frac{1}{s_T}\left[\mathbf{max}_{1\leq k\leq T}\sum_{j=1}^{k}\left(y_j - \bar{y}\right) - \mathbf{min}_{1\leq k\leq T}\sum_{j=1}^{k}\left(y_j - \bar{y}\right)\right] \tag{3.10}$$

The term within third bracket shows the range of partial sum of deviations, and it is rescaled by dividing with s_T, the standard deviation, and $\bar{y}$ is the mean. The R/S statistic is quite robust in itself. It has the ability to detect long range dependence in non-Gaussian time series with large skewness and kurtosis. Traditional methods such as serial autocorrelation can be used to locate long range dependence only

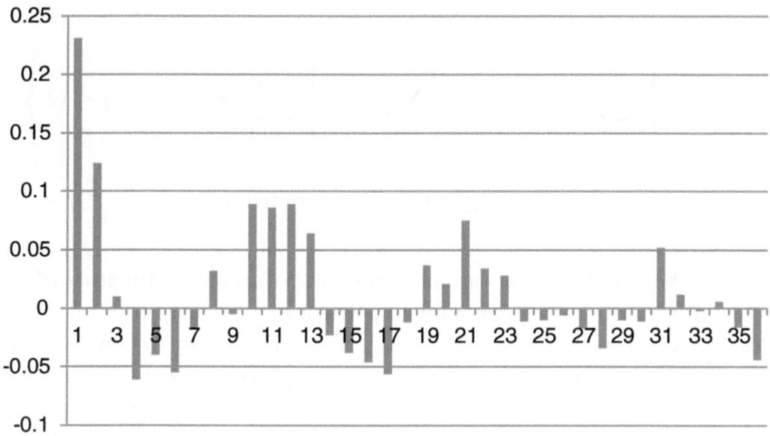

Fig. 3.5 Autocorrelation function for gray portfolio (36 lags)

for models with near Gaussian effects. However, this method has severe shortcomings when used for non-Gaussian time series. According to Mandelbrot and Wallis (1969), "when an ACF analysis program is used blindly for such t.s., the degree of dependence is grossly underrated." Further, R/S analysis has its advantage over other measures of long run dependency such as the variance time analysis. For stochastic processes with infinite variances, the variance time analysis becomes inapplicable. This problem can be overcome by using the R/S analysis. The spectral analysis also becomes inappropriate for the typical economic time series due to its inability to capture the non-periodicity. Again, R/S analysis can be used for such time series.

Nevertheless, the original R/S statistic has also been criticized on many counts. The classical R/S test has been proven to be too weak, i.e., it tends to indicate that a time series has long memory when it does not really have so. Annis and Lloyd (1976) pointed out the small sample bias in the test. Lo (1991) criticized the R/S statistic for being unable to distinguish between short-term and long-term dependence and called it a "severe shortcoming in the applications of the R/S analysis." This observation was backed strongly by Lo and McKinlay's studies (1988, 1990) that showed significant short-range dependence in stock returns. Lo modified the R/S statistics by replacing the denominator, which now is the square root of a consistent estimator of the partial sum's variance.

The modified R/S statistic is defined as

$$\bar{Q}_T = \frac{1}{\hat{\sigma}_T(q)}\left[\mathbf{max}_{1\le k\le T}\sum_{j=1}^{k}(y_j-\bar{y}) - \mathbf{min}_{1\le k\le T}\sum_{j=1}^{k}(y_j-\bar{y})\right] \tag{3.11}$$

where

$$\hat{\sigma}_T^2(q) \equiv \frac{1}{T}\sum_{j=1}^{q}\left(y_j-\bar{y}\right)^2+\frac{2}{T}\sum_{j=1}^{q}\omega_j(q)\left\{\sum_{i=j+1}^{T}\left(y_i-\bar{y}\right)\left(y_{t-j}-\bar{y}\right)\right\}$$

$$=\hat{\sigma}_y^2+2\sum_{i=1}^{q}\omega_j(q)\hat{\gamma}_j \text{ where } \qquad \omega_j(q)\equiv 1-\frac{j}{q+1};\, q<T$$

$\hat{\sigma}_y^2$ and $\hat{\gamma}_j$ are the usual sample variance and autocovariance estimators of y. $\hat{Q}_T$ is different from Q_T only in the denominator, which is the square root of a consistent estimator of the variance of the partial sum. The estimator $\hat{\sigma}_T(q)$ involves not only sums of squared deviations of y_j, but also its weighted autocovariances up to lag q. The weights $\omega_j(q)$ are those suggested by Newey and West (1994) and lead to a positive value of $\hat{\sigma}_T^2(q)$ an estimator of 2π times the (unnormalized) spectral density function of y_t at frequency zero using a Bartlett window.

$\hat{\sigma}_T(q)$ is consistent under the following conditions, according to Phillips' Theorem 4.2 (1987):

A. $\sup_t E[|\varepsilon_t|^{2\beta}] < \infty$ for some $\beta > 2$

B. As T increases without bound, q also increases without bound such that $q \sim o(T^{1/4})$

Lo's modified R/S test has been criticized for being too stringent. It has been shown numerically that even for a synthetic long-memory time series with a moderate value of the Hurst coefficient, the Lo test cannot reject the null hypothesis of short-range dependence.

The rate of growth of the R/S statistic as Hurst found it was close to n^{74}, where n is the number of observations. This phenomenon was named *Hurst Phenomenon*. Moran (1964) tried to explain this phenomenon by incorporating the assumption of an infinite variance. But as pointed out by Mandelbrot (1965) and Mandelbrot and Wallis (1968), assumption of infinite variance is not able to explain the Hurst phenomenon. For that, they suggested a model with a slowly decaying rate of autocorrelation.

The R/S statistic for a process with no long memory increases at a rate $n^{0.5}$. This follows the rule proposed by Mandelbrot (1975) that a process with long memory converges toward a random variable at a rate n^H where H is the Hurst coefficient. For a process with no long memory, $H = 0.5$ and for a long memory process, the value of H lies between 0.5 and 1.

Long memory is said to exist in the market if there is significant autocorrelation between widely separated observations. It is a process in which the autocorrelation function decays asymptotically as a function of the time lag. As a result of the presence of long-term memory, present outcomes can be significantly affected by the outcomes from distant past (Sen 2012).

Table 3.1 Hurst coefficients

Portfolio	Green	Green25	Green50	Green75	Gray
	0.6630	0.4844	0.4297	0.4795	0.5870

Armed with an understanding of the long memory concept, the study uses the R/S test to find out the Hurst coefficient for all five series. The results are summarized in Table 3.1).

The Hurst coefficients confirm the visual evidence. Only 100 % green and 100 % gray portfolios are characterized by long-term memory, not the hybrid portfolios. This finding leads to the automatic next step. If all portfolios are not characterized by long-term memory, will there be a trading rule for all of them? The next section attempts to answer that.

3.3.2.3 Construction of Trading Rule

Step 1. The moving averages generated are all possible combinations of short run, medium run, and long run signals. Fifty-five buying signals are also generated. The buying signals are a series of dummy variables which takes a value of 1 when the price is above some specified threshold and zero, when it is below. The buying signals generated are:

1. Buy3 = price > Mov3, i.e., when the return is higher than the 3 days moving average
2. Buy7 = price > Mov7, i.e., when the return is higher than the 7 days moving average
3. Buy14 = price > Mov14, i.e., when the return is higher than the 14 days moving average
4. Buy21 = price > Mov21, i.e., when the return is higher than the 21 days moving average
5. Buy30 = price > Mov30, i.e., when the return is higher than the 30 days moving average
6. Buy60 = price > Mov60, i.e., when the return is higher than the 60 days moving average
7. Buy100 = price > Mov100, i.e., when the return is higher than the 100 days moving average
8. Buy150 = price > Mov150, i.e., when the return is higher than the 150 days moving average
9. Buy200 = price > Mov200, i.e., when the return is higher than the 200 days moving average
10. Buy270 = price > Mov270, i.e., when the return is higher than the 270 days moving average
11. Buy37 = Mov3 > Mov7, i.e., when the 3 days moving average is higher than the 7 days moving average
12. Buy314 = Mov3 > Mov14, i.e., when the 3 days moving average is higher than the 14 days moving average

13. Buy321 = Mov3 > Mov21, i.e., when the 3 days moving average is higher than the 21 days moving average
14. Buy330 = Mov3 > Mov30, i.e., when the 3 days moving average is higher than the 30 days moving average
15. Buy360 = Mov3 > Mov60, i.e., when the 3 days moving average is higher than the 60 days moving average
16. Buy3100 = Mov3 > Mov100, i.e., when the 3 days moving average is higher than the 100 days moving average
17. Buy3150 = Mov3 > Mov150, i.e., when the 3 days moving average is higher than the 150 days moving average
18. Buy3200 = Mov3 > Mov200, i.e., when the 3 days moving average is higher than the 200 days moving average
19. Buy3270 = Mov3 > Mov270, i.e., when the 3 days moving average is higher than the 270 days moving average
20. Buy714 = Mov7 > Mov14, i.e., when the 7 days moving average is higher than the 14 days moving average
21. Buy721 = Mov7 > Mov21, i.e., when the 7 days moving average is higher than the 21 days moving average
22. Buy730 = Mov7 > Mov30, i.e., when the 7 days moving average is higher than the 30 days moving average
23. Buy760 = Mov7 > Mov60, i.e., when the 7 days moving average is higher than the 60 days moving average
24. Buy7100 = Mov7 > Mov100, i.e., when the 7 days moving average is higher than the 100 days moving average
25. Buy7150 = Mov7 > Mov150, i.e., when the 7 days moving average is higher than the 150 days moving average
26. Buy7200 = Mov7 > Mov200, i.e., when the 7 days moving average is higher than the 200 days moving average
27. Buy7270 = Mov7 > Mov270, i.e., when the 7 days moving average is higher than the 270 days moving average
28. Buy1421 = Mov14 > Mov21, i.e., when the 14 days moving average is higher than the 21 days moving average
29. Buy1430 = Mov14 > Mov30, i.e., when the 14 days moving average is higher than the 30 days moving average
30. Buy1460 = Mov14 > Mov60, i.e., when the 14 days moving average is higher than the 60 days moving average
31. Buy14100 = Mov14 > Mov100, i.e., when the 14 days moving average is higher than the 100 days moving average
32. Buy14150 = Mov14 > Mov150, i.e., when the 14 days moving average is higher than the 150 days moving average
33. Buy14200 = Mov14 > Mov200, i.e., when the 14 days moving average is higher than the 200 days moving average
34. Buy14270 = Mov14 > Mov270, i.e., when the 14 days moving average is higher than the 270 days moving average

35. Buy2130 = Mov21 > Mov30, i.e., when the 21 days moving average is higher than the 30 days moving average
36. Buy2160 = Mov21 > Mov60, i.e., when the 21 days moving average is higher than the 60 days moving average
37. Buy21100 = Mov21 > Mov100, i.e., when the 21 days moving average is higher than the 100 days moving average
38. Buy21150 = Mov21 > Mov150, i.e., when the 21 days moving average is higher than the 150 days moving average
39. Buy21200 = Mov21 > Mov200, i.e., when the 21 days moving average is higher than the 200 days moving average
40. Buy21270 = Mov21 > Mov270, i.e., when the 21 days moving average is higher than the 270 days moving average
41. Buy3060 = Mov30 > Mov60, i.e., when the 30 days moving average is higher than the 60 days moving average
42. Buy30100 = Mov30 > Mov100, i.e., when the 30 days moving average is higher than the 100 days moving average
43. Buy30150 = Mov30 > Mov150, i.e., when the 30 days moving average is higher than the 150 days moving average
44. Buy30200 = Mov30 > Mov200, i.e., when the 30 days moving average is higher than the 200 days moving average
45. Buy30270 = Mov30 > Mov270, i.e., when the 30 days moving average is higher than the 270 days moving average
46. Buy60100 = Mov60 > Mov100, i.e., when the 60 days moving average is higher than the 100 days moving average
47. Buy60150 = Mov60 > Mov150, i.e., when the 60 days moving average is higher than the 150 days moving average
48. Buy60200 = Mov60 > Mov200, i.e., when the 60 days moving average is higher than the 200 days moving average
49. Buy60270 = Mov60 > Mov270, i.e., when the 60 days moving average is higher than the 270 days moving average
50. Buy100150 = Mov100 > Mov150, i.e., when the 100 days moving average is higher than the 150 days moving average
51. Buy100200 = Mov100 > Mov200, i.e., when the 100 days moving average is higher than the 200 days moving average
52. Buy100270 = Mov100 > Mov270, i.e., when the 100 days moving average is higher than the 270 days moving average
53. Buy150200 = Mov150 > Mov200, i.e., when the 150 days moving average is higher than the 200 days moving average
54. Buy150270 = Mov150 > Mov270, i.e., when the 150 days moving average is higher than the 270 days moving average
55. Buy200270 = Mov200 > Mov270, i.e., when the 200 days moving average is higher than the 270 days moving average

Once the moving averages are constructed,

Step 2. The daily return of the portfolio is regressed upon a constant. This equation will serve as the general buy and sell strategy and will work as a benchmark in determining the profitability of a trading rule.

Step 3. In the next step, lagging the buy rule signals by one period, 55 regression equations are generated (One regression for one moving average) for each portfolio. In the following equations, Y_i represents each of the five portfolios. Each regression is estimated and the regression that gives highest daily return, i.e., the coefficient a is considered. The regression returning highest coefficient will be regarded as the best trading rule.

$$
\begin{array}{lll}
Y_i = c + a_1\text{buy}3(-1) & Y_i = c + a_2\text{buy}7(-1) & Y_i = c + a_3\text{buy}14(-1) \\
Y_i = c + a_4\text{buy}21(-1) & Y_i = c + a_5\text{buy}30(-1) & Y_i = c + a_6\text{buy}60(-1) \\
Y_i = c + a_7\text{buy}100(-1) & Y_i = c + a_8\text{buy}150(-1) & Y_i = c + a_9\text{buy}200(-1) \\
Y_i = c + a_{10}\text{buy}270(-1) & Y_i = c + a_{11}\text{buy}37(-1) & Y_i = c + a_{12}\text{buy}314(-1) \\
Y_i = c + a_{13}\text{buy}321(-1) & Y_i = c + a_{14}\text{buy}330(-1) & Y_i = c + a_{15}\text{buy}360(-1) \\
Y_i = c + a_{16}\text{buy}3100(-1) & Y_i = c + a_{17}\text{buy}3150(-1) & Y_i = c + a_{18}\text{buy}3200(-1) \\
Y_i = c + a_{19}\text{buy}3270(-1) & Y_i = c + a_{20}\text{buy}714(-1) & Y_i = c + a_{21}\text{buy}721(-1) \\
Y_i = c + a_{22}\text{buy}730(-1) & Y_i = c + a_{23}\text{buy}760(-1) & Y_i = c + a_{24}\text{buy}7100(-1) \\
Y_i = c + a_{25}\text{buy}7150(-1) & Y_i = c + a_{26}\text{buy}7200(-1) & Y_i = c + a_{27}\text{buy}7270(-1) \\
Y_i = c + a_{28}\text{buy}1421(-1) & Y_i = c + a_{29}\text{buy}1430(-1) & Y_i = c + a_{30}\text{buy}1460(-1) \\
Y_i = c + a_{31}\text{buy}14100(-1) & Y_i = c + a_{32}\text{buy}14150(-1) & Y_i = c + a_{33}\text{buy}14200(-1) \\
Y_i = c + a_{34}\text{buy}14270(-1) & Y_i = c + a_{35}\text{buy}2130(-1) & Y_i = c + a_{36}\text{buy}2160(-1) \\
Y_i = c + a_{37}\text{buy}21100(-1) & Y_i = c + a_{38}\text{buy}21150(-1) & Y_i = c + a_{39}\text{buy}21200(-1) \\
Y_i = c + a_{40}\text{buy}21270(-1) & Y_i = c + a_{41}\text{buy}3060(-1) & Y_i = c + a_{42}\text{buy}30100(-1) \\
Y_i = c + a_{43}\text{buy}30150(-1) & Y_i = c + a_{44}\text{buy}30200(-1) & Y_i = c + a_{45}\text{buy}30270(-1) \\
Y_i = c + a_{46}\text{buy}60100(-1) & Y_i = c + a_{47}\text{buy}60150(-1) & Y_i = c + a_{48}\text{buy}60200(-1) \\
Y_i = c + a_{49}\text{buy}60270(-1) & Y_i = c + a_{50}\text{buy}100150(-1) & Y_i = c + a_{51}\text{buy}100200(-1) \\
Y_i = c + a_{52}\text{buy}100270(-1) & Y_i = c + a_{53}\text{buy}150200(-1) & Y_i = c + a_{54}\text{buy}150270(-1) \\
Y_i = c + a_{55}\text{buy}200270(-1) & &
\end{array}
$$

Portfolio 1: 100 % Green

The daily return of the green portfolio is regressed upon a constant and the result is summarized in Table 3.2.

From the 55 regressions, GREEN $= c + a_{44}\text{buy}3060(-1)$ gives the highest daily return. Comparing the coefficient in the general buy and sell strategy, (Table 3.2) and the coefficient in the trading rule (Table 3.3), it can be seen that the coefficient of the general strategy is 0.012214, while the coefficient of the trading strategy is 0.036. Clearly, the coefficient of the trading rule is significantly higher. This implies, when the investor is following a trading rule such that he buys whenever the 30 day moving average is greater than the 60 day moving average and sells whenever the 30 day moving average falls below the 60 day moving average, he is getting a daily return of 0.02 %. The daily return looks small, but is significant (at 1 % level). However, if the investor is following a general buy

Table 3.2 Regression result of green on a constant (general buy and sell strategy)

Dependent variable: green				
Variable	Coefficient	Std. error	*t*-statistic	Prob.
C	0.012214	0.000461	26.52174	0

Table 3.3 Regression result of green based on the trading rule

Variable	Coefficient	Std. Error	*t*-Statistic	Prob.	R-squared	Adjusted R-squared
C	0.010284	0.000701	14.66988	0	0.020369	0.018838
BUY3060(−-1)	0.0036	0.000987	3.647869	0.0003		

and sell strategy, he will be earning a lower profit of 0.0036 % daily. Therefore, a 30–60 day moving average trading strategy is a significantly better strategy.

Portfolio 2: 25 % Green (G25)

The daily return of the 25 % green (G25) is regressed upon a constant and the result is summarized in Table 3.4.

From the 55 regressions, $G25 = c + a_{44}\text{buy}150200(-1)$ gives the highest daily return of 0.001618 %. The result of the regression is summarized in Table 3.5. The coefficient of BUY150200(−1) is 0.005, while the coefficient of the general buy and sell strategy is 0.00085. However, the daily return for the trading rule is not significant even at a 10 % level, whereas the daily return in the general strategy is significant at a strong 1 % level. Therefore, it can be inferred that a general buy and sell strategy will prove more beneficial for this portfolio than a trading rule.

Portfolio 33: 50 % Green

The daily return of G50 is regressed upon a constant and the result is summarized in Table 3.6.

From the 55 regressions, $G50 = c + a_{44}\text{buy}150270(-1)$ gives the highest daily return. The coefficient of buy150270(−1) is 0.0009, and the intercept for the general buy and sell strategy is 0.000543. The return from the trading rule, however, is NOT statistically significant. Therefore, it can be said that for this portfolio, there is no significant trading rule. Therefore, the investor will be making a higher return by following the general rule (Table 3.7).

Portfolio 4: 75 % Green

The daily return of G75 is regressed upon a constant and the result is summarized in Table 3.8.

From the 55 regressions, $G75 = c + a_{48}\text{buy}150270(-1)$ gives the highest daily return. The coefficient of BUY150270(−1) is 0.00099. This means, when the

Table 3.4 Regression result of 25 % green (G25) on a constant (general buy and sell strategy)

Dependent variable: G25						
Variable	Coefficient	Std. error	*t*-statistic	Prob.	R-squared	Adjusted R^2
C	0.000574	0.000223	2.573593	0.010	0.000	0.000

Table 3.5 Regression result of 25 % green (G25) based on the trading rule

Dependent variable: G25						
Variable	Coefficient	Std. error	*t*-statistic	Prob.	R^2	Adjusted R^2
C	0.000155	0.000382	0.406533	0.6845	0.0047	0.0027
BUY150200(−1)	0.00085	0.000553	1.536426	0.1251		

Table 3.6 Regression result of 50 % green (G50) on a constant (general buy and sell strategy)

Dependent variable: G50						
Variable	Coefficient	Std. error	*t*-statistic	Prob.	R^2	Adjusted R^2
C	0.000543	0.000239	2.273588	0.0233	0.000	0.000

Table 3.7 Regression result of 50 % green (G50) based on the trading rule

Dependent variable: G50						
Variable	Coefficient	Std. error	*t*-statistic	Prob.	R^2	Adjusted R^2
C	0.000293	0.000368	0.79541	0.4268	0.0039	0.001583
BUY150270(−1)	0.000918	0.000708	1.297469	0.1952		

Table 3.8 Regression result of 75 % green (G75) on a constant (general buy and sell strategy)

Dependent variable: G75						
Variable	Coefficient	Std. error	*t*-statistic	Prob.	R^2	Adjusted R^2
C	0.00005	0.0003	0.1946	0.8458	0.000	0.000

investor is following the general buy and sell strategy, he is getting a daily return of 0.00005 %, and he will be getting a higher return of 0.00099 % when he follows the trading rule. The trading rule return, however, is not significant at 10 % level. Therefore, it can be said that the investors are better off following the general buy and sell strategy for this portfolio (Table 3.9).

Portfolio 5: 100 % Gray

The daily return of GRAY is regressed upon a constant and the result is summarized in Table 3.10.

Table 3.9 Regression result of 75 % green (G75) based on the trading rule

Dependent variable: G75						
Variable	Coefficient	Std. error	*t*-statistic	Prob.	R^2	Adjusted R^2
C	−0.0002	0.0004	−0.4417	0.6589	0.00409	0.001774
BUY150270(−1)	0.00099	0.0007	1.3288	0.1846		

Table 3.10 Regression result of 100 % gray on a constant (general buy and sell strategy)

Dependent variable: gray						
Variable	Coefficient	Std. error	*t*-statistic	Prob.	R^2	Adjusted R^2
C	−0.000147	0.000295	−0.496284	0.6198	0.000	0.000

Table 3.11 Regression result of 100 % gray based on the trading rule

Dependent variable: gray						
Variable	Coefficient	Std. error	*t*-statistic	Prob.	R^2	Adjusted R^2
C	−0.00104	0.000465	−2.23622	0.0258	0.006282	0.003971
BUY21270(−1)	0.00136	0.000825	1.648742	0.0999		

From the 55 regressions, $\text{GRAY} = c + a_{44}\text{buy21270}(-1)$ gives the highest daily return. The coefficient of BUY21270(−1) is 0.00136 and the intercept in the general buy and sell strategy is −0.000147. This means, when the investor is following the trading rule of buying whenever the 21 day moving is higher than the 270 days moving average and do not hold whenever it is below the 270 days moving average, he is getting a daily return of 0.00136 %, and he will be making a loss of 0.000147 % when he is following the general strategy. The trading rule return is statistically significant at 5 % level as well. Therefore, there exists an optimum trading rule for this portfolio (Table 3.11).

Looking at the results above from the study in context of the five portfolio, one clear pattern emerges. For the "hybrid" portfolios, the general buy and sell strategies proved to be a better strategy, while for the "pure" portfolios, a trading rule leads to higher return. Now, if the 100 % green and the 100 % gray portfolios are compared, it can be seen that the trading rule in the green portfolio leads to a 0.0036 % daily return, while the trading rule in the gray portfolio leads to a lower 0.00136 % daily return. This reinforces the argument for green assets once again.

3.4 Does Green Really Rule the Others? A Bird's Eye Perspective

This section investigated whether a "greener" portfolio leads itself to a higher than expected return based on a technical momentum rule. The market has generally been efficient, therefore the likelihood of garnering a higher than average return consistently is a possibility. Close examination of the data revealed that apart from the two"pure" portfolios, i.e., the 100 % green and the 100 % gray portfolios, no other "hybrid" portfolio is characterized by long-term memory. Also, a deeper analysis based on moving average-based trading rule revealed that these two "pure" portfolios are also characterized by long run momentum. In these two cases, the trading rule clearly yields a higher return than the general buy and sell strategy. The hybrid strategies, regardless of the percentage of green stocks in them, show no evidence of a trading rule. Also, a closer inspection reveals that the 100 % green portfolio has a higher daily return than a 100 % gray portfolio when the optimum trading rule is followed. Several studies have tried to provide an explanation for this higher than average momentum return. While some studies attribute this higher return to higher risk, which was previously not captured in traditional models (Jegadeesh and Titman 2001, 2002), some studies attribute the higher risk to firm specific informations (Berk et al. 1999; Chordia and Shivakumar 2002) and expected growth rates (Johnson 2002). Behavioral theorists have attempted to explain the momentum return with a perspective bias and henceforth over or under-reaction on part of investors (Jegadeesh and Titman 2001, 2002, 2005; Frazzini 2006) and even on a "self-attribution bias" (Daniel et al. 1998).

Therefore, in short, it can rightly be said, that being inefficient, this market lends itself significantly to predictions and it is possible to have a higher than average return if the right trading rule can be estimated. More importantly, a green investment indeed leads to higher returns.

References

Asness CS, Moskowitz TJ, Pedersen LH (2009) Value and momentum everywhere, AFA 2010 Atlanta meetings paper. Available at SSRN: http://ssrn.com/abstract=1363476, Accessed 6 Mar 2009

Alexander S (1961) Price movements in speculative markets: trends or random walks. Ind Manage Rev 2(2):7–26

Annis AA, Lloyd EH (1976) The expected value of the adjusted rescaled hurst range of independent normal summands. Biometrika 63:111–116

Berk JB, Green RC, Naik V (1999) Optimal investment, growth options, and security returns. J Finance 54(5):1153–1608

Bessembinder H, Chan K (1995) The profitability of technical trading rules in the asian stock markets. Pac-Basin Financ J 3(2–3):257–284

Brock W, Lakonishok J, LeBaron B (1992) Simple technical trading rules and the stochastic properties of stock returns. J Finance 47:1731–1764

Carhart MM (1997) On persistence in mutual fund performance. J Finance 52(1):57–82

Charoenwong BG (2012) An exploration of simple optimized technical trading strategies and their trading costs, University of Chicago, Booth School of Business, Available online at: http://deepblue.lib.umich.edu/bitstream/handle/2027.42/91813/chben.pdf, Accessed on: 16 Dec 2013

Chordia T, Shivakumar L (2002) Momentum, business cycle and time-varying expected returns. J Finance 57(2):985–1019

Conrad J, Kaul G (1998) An anatomy of trading strategies. Rev of Fin Stud 11:489–519

Daniel KD, Hirshleifer D, Subrahmanyam A (1998) A theory of overconfidence, self-attribution, and security under and overreactions. J Finance 53(6):1839–1885

Fama E, Blume M (1966) Filter rules and stock-market trading. J Bus 39(1):226–241

Fama EF, French KR (1988) Permanent and temporary components of stock prices. J Polit Econ 96:246–273
Fama EF, French KR (1993) Common risk factors in the returns on stocks and bonds. J Financ Econ 33(1):3–56
Frazzini A (2006) The disposition effect and underreaction to news. J Finance 61(4):2017–2046
Hameed A, Yuanto K (2002) Momentum strategies: evidence from the pacific basin stock markets. J Financ Res 25(3):383–397
Harvey C (1955) The cross-section of volatility and autocorrelation in emerging markets. Finanzmarkt und Portfolio Management 9(1995a):12–34
Hudson R, Dempsey M, Keasey K (1996) A note on the weak form efficiency of capital markets: the application of simple technical trading rules to UK stock prices—1935 to 1994. J Bank Financ 20(1996):1121–1132
Hurst H (1951) Long term storage capacity of reservoirs. Trans Am Soc Civ Eng 116:770–799
Jegadeesh N (1990) Evidence of predictable behaviour of security returns. J Finance 45:881–898
Jegadeesh N, Titman S (1993) Returns to buying winners and selling losers: implications for stock market efficiency. J Finance 48:65–91
Jegadeesh N, Titman S (1995) Overreaction, delayed reaction and contrarian profits. Rev Financ Stud 8:973–993
Jegadeesh N, Titman S (2001) Profitability of momentum strategies: an evaluation of alternative explanations. J Finance 56(2):699–720
Jegadeesh N, Titman S (2002) Cross-sectional and time-series determinants of momentum returns. Rev Financ Stud 15(1):143–157
Jegadeesh N, Titman S (2005) Momentum. In: Thaler R (ed) Advances in behavioral finance, II. Princeton Press, Princeton
Johansen AEB, Villaddsen M (2013) Time-series momentum: an empirical analysis of performance and option-like behavior, Available online at: http://pure.au.dk/portal-asb-student/files/55325357/Thesis.pdf, Accessed on: 17 Dec 2013
Johnson TC (2002) Rational momentum effects. J Finance 57:585–608
Lee CMC, Swaminathan B (2000) Price momentum and trading volume. J. Finance 55:2017–2069
Lewellen J (2002) Momentum and autocorrelation in stock returns. Rev Finan Stud 15(2):533–563
Liu W, Norman S, Xu X (1999) UK momentum tests. J Bus Financ Account 26(9/10):1043–1091
Lo, AW (1991) Long-term memory in stock market prices. Econometrica 59:1279–1313
Lo AW, MacKinlay CA (1988) Stock prices do not follow random walks: evidence from a simple specification test. Rev Financ Stud 1(1):41–66
Lo AW, Mamaysky H, Wang J (2000) Foundations of technical analysis: computationalalgorithms, statistical inference, and empirical implementation. J Finance 55:1705–1765
Mandelbrot B (1965) Forecasts of future prices, unbiased markets, and 'martingale' models. J Bus 39:242–255
Mandelbrot BB (1972) Statistical methodology for non-periodic cycles; from the covariance to R/S analysis. Ann Econ Soc Meas 1:259–260
Mandelbrot BB (1975) A fast fractional gaussian noise generator. Water Resour Res 7:543–553
Mandelbrot B, Wallis J (1968) Noah, Joseph and operational hydrology. Water Resour Res 4:909–918
Mandelbrot B, Wallis J (1969) Computer experiments with fractional gaussian noises: parts 1, 2, 3. Water Resour Res 5:228–267
Mashaushi KRS (2006) An analysis of technical trading strategies, The University of Leeds, Leeds University Business School, http://etheses.whiterose.ac.uk/696/1/uk_bl_ethos_431997.pdf, Accessed 16 Dec 2013
Moran PAP (1964) On the range of cumulative sums. Ann Inst Stat Math 16:109–112
Moskowitz TJ, Ooi YH, Pedersen LH (2012) Time-series momentum. J Financ Econ 104:228–250
Neely CJ, Rapach DE, Tu J, Zhou G (2013) Forecasting the equity risk premium: the role of technical indicators. Manage Sci, forthcoming
Newey, KW, West KD (1994) Automatic lag selection in covariance matrix estimation. Rev Econ Stud 61(4):631–53

Person J (2007) Moving average formula and strategy guide. Daniels trading professional trader series, Available files.meetup.com/1701672/Moving_Average_Guide.pdf, Accessed 16 Dec 2013
Ready M (1997) Profits from technical trading rules. University of Wisconsin-Madison working paper
Richards AJ (1997) Winner–loser reversals in national stock market indices: can they be explained? J Finance 52:2129–2144
Rouwenhorst KG (1998) International momentum strategies. J Finance 53:267–284
Sen C (2012) Indian foreign exchange market: a study into volatility and regime switch. LAP LAMBERT Academic Publishing, Germany

Chapter 4
Epilogue

Abstract With rapidly rising global temperature, containment of carbon footprint is an issue of paramount importance. This study shows that green portfolios are more stable choice of investment over gray portfolios, especially during the times of crisis. For major global firms, their profitability is closely connected to a sustainable production process. With an increasing number of Indian firms adapting green policies, it can be said the progress in the right direction has started.

Keywords Carbon footprint · Negative externality · Green technology · Sustainable production

As we have moved into the second decade of twenty-first century, the danger of global warming and environmental degradation is looming bigger by day. Bigger carbon footprints impose a negative externality on the economy, which makes a significant diversion between the private costs borne by the firm and the social cost borne by the economy. Therefore, from the economy's perspective, green technologies are more desired. However, the green technologies will only be desired to the firms if it ensures a higher profit. Once this is ensured, the social and private choices will automatically be aligned without any government intervention and therefore deadweight loss.

To emphasize the investment-worthiness of green and nongreen assets based on their performance, this study compared five portfolios consisting of 100 % green, 75 % green-25 % gray, 50 % green-50 % gray, 25 % green-75 % gray, and 100 % gray stocks. As the study reveals, the green portfolios turn out to be the global minimum variance portfolio, and they dominate the gray in terms of the own risk as well as the market risk. The green portfolios are also inherently more stable in nature. Not only that the green portfolios have a higher probability of surviving a financial crisis, but also making it a better investment choice during a volatile period. The study delved deeper and farther revealed the existence of a long run moving average trading rule for the pure green portfolio resulting in a higher daily return than gray portfolio, making green an automatic investment choice over the gray.

G. Chakrabarti and C. Sen, *Green Investing*, SpringerBriefs in Finance,
DOI 10.1007/978-81-322-2026-8_4

Hence, following less-carbon investment path is the most rational and obvious choice for the investors in the Indian market. The findings of this paper reinforce the fact that in the coming decades, in order to be profitable and attract more investors, the companies need to adapt more green technologies. Fortunately, some of the best companies in the world have already started adapting green production policies into their best practices. For example, IKEA, one of the biggest furniture manufacturers in the world with €27.628 billion revenue in year 2012,[1] has been using sustainable production process for a very long time. A large furniture manufacturer such as IKEA is supposed to have a significant environmental footprint, especially on the forestry. However, the company is an early adapter of sustainable forestry practices, planting more trees than are used by them in the production. Not only that, IKEA encourages the use of more energy saving and eco-friendly LED and halogen lamps, rather than the traditional and less eco-friendly incandescent ones. Not only that, many IKEA stores in the USA have installed solar panels in order to meet their energy need and lower the carbon footprint. Timberland, a boot and outerwear manufacturing company, has reduced their carbon emissions by 38 % since 2006 by making their stores more energy efficient (www.responsibility.timberland.com) by installing LED lamps. The company aims to reduce its impact on the environment by minimizing the use of harmful chemicals and resource consumption toward the making of its products. Also, Timberland follows a unique practice of providing the green product index for every product manufactured, which summarizes the greenhouse gas emission, use of harmful chemicals, resource consumption, and impact on environment made, while manufacturing that product. These two are not the only examples of multinationals that are gradually shifting more toward greener production processes. The number is swelling gradually. Johnson and Johnson in the USA uses 52 % of its power from solar energy, EarthTec is making clothing items out of recyclable plastic bottles, United Airlines has experimented with algae-based biofuels to fly its jets, while Alaskan Air has been flying on biofuels, saving massive energy (Steil 2013), Philips invested billions of euros in green research and Dell has been successfully moving toward its target of reducing its carbon footprint by 40 % by 2015 (www.planetsave.com). And these are to name a few.

The companies discussed above who are adapting green technologies in a big way are all global leaders in their respective areas. The only reason behind it is that the importance of sustainable production process and how closely it is intertwined with the profitability of the company has been realized and as mentioned earlier, the gap between social and private costs imposed by these companies are closing.

According to 2012, Newsweek Green Rankings Report, the Indian company Wipro is global number two, with an impressive 85.4 green score.[2] The importance of going green has been, however, realized and more and more Indian firms are adapting their policies accordingly. It is only imperative for Indian firms to do so, as most Indian metros are ranked very high when it comes to air pollution

---

[1] *Source* http://en.wikipedia.org/wiki/IKEA.

[2] *Source* http://www.newsweek.com/2012/10/22/newsweek-green-rankings-2012-global-500-list.html.

compared to American, European, and even Chinese cities. Adaption of sustainable technologies will not only be ethical and financially prudent, but also it must be the only option before it is too late.

Reference

Steil J (2013) Go green: 15 companies you can feel good about supporting. Available http://www.ivillage.com/15-best-green-companies/7-b-435283#435605. Accessed 27 Dec 2013

Appendix

Table A.1 (Part A)

Dependent variable: probability of crisis (own)				
Method: ML—binary probit (quadratic hill climbing)				
Included observations: 47 after adjustments				
Convergence achieved after 6 iterations				
Covariance matrix computed using second derivatives				
Variable	Coefficient	Std. error	z-statistic	Prob.
C	7.507594	2.659024	2.823439	0.0048
Asset turnover	−0.266978	0.212592	−1.255823	0.2092
Financial charge cover ratio	0.000545	0.000386	1.413019	0.1577
Conditional correlation with market	−0.475697	6.741296	−0.070565	0.9437
Current ratio	0.736209	0.329633	2.233423	0.0255
Debt-equity ratio	0.003866	0.008454	0.457288	0.6475
DPS	−0.043960	0.027033	−1.626167	0.1039
EPS growth	0.005130	0.011096	0.462358	0.6438
Green	−1.85942	0.845388	−2.19948	0.0278
NPM (%)	−0.284073	0.091929	−3.090133	0.0020
RAR	0.588333	0.606710	0.969711	0.3322
Retention ratio	−0.065239	0.026809	−2.433488	0.0150
McFadden R-squared	0.503664	Mean dependent var		0.531915
S.D. dependent var	0.504375	S.E. of regression		0.384303
Akaike info criterion	1.196683	Sum squared resid		5.169107
Schwarz criterion	1.669061	Log likelihood		−16.12204
Hannan–Quinn criterion	1.374442	Deviance		32.24409
Restr. deviance	64.96422	Restr. log likelihood		−32.48211
LR statistic	32.72013	Avg. log likelihood		−0.343022
Prob (LR statistic)	0.000584			
Obs with Dep = 0	22	Total obs		47
Obs with Dep = 1	25			

G. Chakrabarti and C. Sen, *Green Investing*, SpringerBriefs in Finance,
DOI 10.1007/978-81-322-2026-8

Table A.1 (Part B)

Expectation-prediction evaluation for binary specification						
Success cutoff: C = 0.5						
	Estimated equation			Constant probability		
	Dep = 0	Dep = 1	Total	Dep = 0	Dep = 1	Total
P(Dep = 1) <= C	19	4	23	0	0	0
P(Dep = 1) > C	3	21	24	22	25	47
Total	22	25	47	22	25	47
Correct	19	21	40	0	25	25
% correct	86.36	84.00	85.11	0.00	100.00	53.19
% incorrect	13.64	16.00	14.89	100.00	0.00	46.81
Total gain[a]	86.36	−16.00	31.91			
Percent gain[b]	86.36	NA	68.18			
	Estimated equation			Constant probability		
	Dep = 0	Dep = 1	Total	Dep = 0	Dep = 1	Total
E(# of Dep = 0)	17.03	5.22	22.25	10.30	11.70	22.00
E(# of Dep = 1)	4.97	19.78	24.75	11.70	13.30	25.00
Total	22.00	25.00	47.00	22.00	25.00	47.00
Correct	17.03	19.78	36.81	10.30	13.30	23.60
% correct	77.42	79.12	78.32	46.81	53.19	50.20
% incorrect	22.58	20.88	21.68	53.19	46.81	49.80
Total gain[a]	30.61	25.93	28.12			
Percent gain[b]	57.54	55.40	56.47			

[a]Change in "% Correct" from default (constant probability) specification
[b]Percent of incorrect (default) prediction corrected by equation

Table A.2 (Part A)

Dependent variable: MKT_STRESS				
Method: ML—binary probit (quadratic hill climbing)				
Sample (adjusted): 1 48				
Included observations: 47 after adjustments				
Convergence achieved after 8 iterations				
Covariance matrix computed using second derivatives				
Variable	Coefficient	Std. error	z-Statistic	Prob.
C	−0.681962	2.006131	−0.339939	0.7339
Asset turnover	−0.126999	0.171827	−0.739108	0.4598
Financial charge cover ratio	3.71E-05	0.000335	0.110754	0.9118
Conditional correlation with market	11.95253	4.405112	2.709751	0.0120
Current ratio	−0.460842	0.229540	−2.007680	0.0447
Debt-equity ratio	0.045549	0.050002	0.910953	0.3623
DPS	0.017464	0.026240	0.665567	0.5057
EPS	−0.003384	0.007230	−0.468042	0.6398
Green	−1.86978	0.845388	−2.19948	0.0278
NPM (%)	−0.054029	0.046926	−1.151362	0.2496
RAR	−1.241923	9.059794	−0.137081	0.8910
Retention ratio	0.003531	0.019701	0.179250	0.8577
McFadden R-squared	0.302974	Mean dependent var		0.361702
S.D. dependent var	0.485688	S.E. of regression		0.446710
Akaike info criterion	1.422894	Sum squared resid		6.984250
Schwarz criterion	1.895272	Log likelihood		−21.43801
Hannan–Quinn criterion	1.600653	Deviance		42.87603
Restr. deviance	61.51278	Restr. log likelihood		−30.75639
LR statistic	18.63675	Avg. log likelihood		−0.456128
Prob (LR statistic)	0.067934			
Obs with Dep = 0	30	Total obs		47
Obs with Dep = 1	17			

Table A.2 (Part B)

Expectation-prediction evaluation for binary specification						
Success cutoff: C = 0.5						
	Estimated equation			Constant probability		
	Dep = 0	Dep = 1	Total	Dep = 0	Dep = 1	Total
P(Dep = 1) <= C	27	8	35	30	17	47
P(Dep = 1) > C	3	9	12	0	0	0
Total	30	17	47	30	17	47
Correct	27	9	36	30	0	30
% correct	90.00	52.94	76.60	100.00	0.00	63.83
% incorrect	10.00	47.06	23.40	0.00	100.00	36.17
Total gain[a]	−10.00	52.94	12.77			
	Estimated equation			Constant probability		
	Dep = 0	Dep = 1	Total	Dep = 0	Dep = 1	Total
E(# of Dep = 0)	23.09	7.26	30.35	19.15	10.85	30.00
E(# of Dep = 1)	6.91	9.74	16.65	10.85	6.15	17.00
Total	30.00	17.00	47.00	30.00	17.00	47.00
Correct	23.09	9.74	32.83	19.15	6.15	25.30
% correct	76.96	57.27	69.84	63.83	36.17	53.83
% incorrect	23.04	42.73	30.16	36.17	63.83	46.17
Total gain[a]	13.13	21.10	16.02			
Percent gain[b]	36.31	33.06	34.69			

[a]Change in "% Correct" from default (constant probability) specification
[b]Percent of incorrect (default) prediction corrected by equation

Table A.3 (Part A)

General buy and sell strategy for 100 % green portfolio				
Variable	Coefficient	Std. error	t-statistic	Prob.
C	0.012214	0.000461	26.52174	0
R-squared	0	Mean dependent var		0.012214
Adjusted R-squared	0	S.D. dependent var		0.012201
S.E. of regression	0.012201	Akaike info criterion		−5.9731
Sum squared resid	0.104362	Schwarz criterion		−5.96661
Log likelihood	2,097.558	Hannan–Quinn criterion		−5.97059
Durbin–Watson statistic	1.737959			

Table A.3 (Part B)

Trading strategy for 100 % green portfolio				
Variable	Coefficient	Std. error	t-statistic	Prob.
C	0.010284	0.000701	14.66988	0
BUY3060(−1)	0.0036	0.000987	3.647869	0.0003
R-squared	0.020369	Mean dependent var		0.012101
Adjusted R-squared	0.018838	S.D. dependent var		0.012621
S.E. of regression	0.012502	Akaike info criterion		−5.9228
Sum squared resid	0.100026	Schwarz criterion		−5.9089
Log likelihood	1,903.22	Hannan–Quinn criterion		−5.91741
F-statistic	13.30695	Durbin–Watson statistic		1.781974
Prob (F-statistic)	0.000286			

Table A.4 (Part A)

General buy and sell strategy for 25 % green portfolio				
Variable	Coefficient	Std. error	t-statistic	Prob.
C	0.000574	0.000223	2.573593	0.0103
R-squared	0	Mean dependent var		0.000574
Adjusted R-squared	0	S.D. dependent var		0.005906
S.E. of regression	0.005906	Akaike info criterion		−7.424322
Sum squared resid	0.02445	Schwarz criterion		−7.417835
Log likelihood	2,606.937	Hannan–Quinn criterion		−7.421815

Table A.4 (Part B)

Trading rule for 25 % green portfolio				
Variable	Coefficient	Std. error	t-statistic	Prob.
C	0.000155	0.000382	0.406533	0.6845
BUY150200(−1)	0.00085	0.000553	1.536426	0.1251
R-squared	0.004699	Mean dependent var		0.000562
Adjusted R-squared	0.002708	S.D. dependent var		0.0062
S.E. of regression	0.006191	Akaike info criterion		−7.327388
Sum squared resid	0.019166	Schwarz criterion		−7.310581
Log likelihood	1,841.174	Hannan–Quinn criterion		−7.320794
F-statistic	2.360605	Durbin–Watson statistic		2.097381
Prob (F-statistic)	0.125066			

Table A.5 (Part A)

General buy and sell strategy for 50 % green portfolio				
Variable	Coefficient	Std. error	t-statistic	Prob.
C	0.000543	0.000239	2.273588	0.0233
R-squared	0	Mean dependent var		0.000543
Adjusted R-squared	0	S.D. dependent var		0.006328
S.E. of regression	0.006328	Akaike info criterion		−7.286211
Sum squared resid	0.028071	Schwarz criterion		−7.279724
Log likelihood	2,558.46	Hannan–Quinn criterion		−7.283704
Durbin–Watson statistic	2.275404			

Table A.5 (Part B)

Trading rule for 50 % green portfolio				
Variable	Coefficient	Std. error	t-statistic	Prob.
C	0.000293	0.000368	0.79541	0.4268
BUY150270(−1)	0.000918	0.000708	1.297469	0.1952
R-squared	0.0039	Mean dependent var		0.000542
Adjusted R-squared	0.001583	S.D. dependent var		0.00654
S.E. of regression	0.006535	Akaike info criterion		−7.21862
Sum squared resid	0.018365	Schwarz criterion		−7.199785
Log likelihood	1,561.222	Hannan–Quinn criterion		−7.211184
F-statistic	1.683426	Durbin–Watson statistic		2.272725
Prob (F-statistic)	0.195165			

Table A.6 (Part A)

General buy and sell strategy for 75 % green portfolio				
Variable	Coefficient	Std. error	t-statistic	Prob.
C	5.00E-05	0.000257	0.19456	0.8458
R-squared	0	Mean dependent var		5.00E-05
Adjusted R-squared	0	S.D. dependent var		0.006813
S.E. of regression	0.006813	Akaike info criterion		−7.138561
Sum squared resid	0.032538	Schwarz criterion		−7.132074
Log likelihood	2,506.635	Hannan–Quinn criterion		−7.136054
Durbin–Watson statistic	2.263722			

Table A.6 (Part B)

Trading rule for 75 % green portfolio				
Variable	Coefficient	Std. error	t-statistic	Prob.
C	−0.0002	0.0004	−0.4417	0.6589
BUY150270(−1)	0.0010	0.0007	1.3288	0.1846
R-squared	0.00409	Mean dependent var		9.68E-05
Adjusted R-squared	0.001774	S.D. dependent var		0.006884
S.E. of regression	0.006877	Akaike info criterion		−7.116521
Sum squared resid	0.020339	Schwarz criterion		−7.097686
Log likelihood	1,539.169	Hannan–Quinn criterion		−7.109085
F-statistic	1.765794	Durbin–Watson statistic		2.327136
Prob (F-statistic)	0.184608			

Table A.7 (Part A)

General buy and sell strategy for 100 % gray portfolio				
Variable	Coefficient	Std. error	t-statistic	Prob.
C	−0.00015	0.000295	−0.49628	0.6198
R-squared	0	Mean dependent var		−0.00015
Adjusted R-squared	0	S.D. dependent var		0.007828
S.E. of regression	0.007828	Akaike info criterion		−6.86086
Sum squared resid	0.042953	Schwarz criterion		−6.85437
Log likelihood	2,409.161	Hannan–Quinn criterion		−6.85835
Durbin–Watson statistic	1.536893			

Table A.7 (Part B)

Trading rule for 100 % gray portfolio				
Variable	Coefficient	Std. error	t-statistic	Prob.
C	−0.00104	0.000465	−2.23622	0.0258
BUY21270(−1)	0.00136	0.000825	1.648742	0.0999
R-squared	0.006282	Mean dependent var		−0.00061
Adjusted R-squared	0.003971	S.D. dependent var		0.007994
S.E. of regression	0.007978	Akaike info criterion		−6.81953
Sum squared resid	0.027372	Schwarz criterion		−6.80069
Log likelihood	1,475.017	Hannan–Quinn criterion		−6.81209
F-statistic	2.71835	Durbin–Watson statistic		1.561883
Prob (F-statistic)	0.099931			

Table A.8 Coefficients of all possible trading rules: 100 % green

	7	14	21	30	60	100	150	200	270
3	0.000791	0.000618	0.001601	0.001762	0.002769	0.003375	0.002659	0.002907	0.002894
7		0.000599	0.001261	0.00184	0.002637	0.003143	0.002333	0.003166	0.003237
14			0.002114	0.002378	0.003518	0.003716	0.003009	0.003013	0.003481
21				0.002861	0.003443	0.003499	0.00239	0.002637	0.003233
30					**0.0036**	0.002639	0.00256	0.00235	0.00297
60						0.001623	0.002706	0.001739	0.001739
100							0.003443	0.002741	0.002682
150								0.001657	0.001396
200									0.001453

Table A.9 Coefficients of all possible trading rules: 75 % green

	7	14	21	30	60	100	150	200	270
3	0.00016	−0.00021	−0.00042	−0.00035	−0.00032	−0.00028	−0.00029	−0.00058	−0.00063
7		−0.00071	−0.00088	−0.00091	−0.00060	−0.00070	−0.00066	−0.00072	−0.00007
14			−0.00088	−0.00067	−0.00094	−0.00051	−0.00050	−0.00065	−0.00047
21				−0.00017	−0.00011	0.00029	0.00040	0.00070	0.00051
30					−0.00032	−0.00003	0.00048	0.00052	0.00028
60						0.00039	0.00023	0.00069	0.00051
100							−0.00013	0.00016	−0.00012
150								**0.00085**	0.00056
200									0.00001

Table A.10 Coefficients of all possible trading rules: 50 % green

	7	14	21	30	60	100	150	200	270
3	0.00028	−0.00046	−0.00075	−0.00054	−0.00034	−0.00035	−0.00043	−0.00048	−0.00092
7		−0.00108	−0.00106	−0.00080	−0.00074	−0.00065	−0.00072	−0.00048	−0.00020
14			−0.00115	−0.00056	−0.00080	−0.00049	−0.00058	−0.00061	−0.00057
21				−0.00061	−0.00018	0.00026	0.00024	0.00079	0.00025
30					−0.00061	−0.00018	0.00036	0.00050	0.00015
60						0.00029	0.00020	0.00056	0.00030
100							0.00012	0.00040	0.00025
150								0.00090	**0.00092**
200									−0.00006

Table A.11 Coefficients of all possible trading rules: 25 % green

	7	14	21	30	60	100	150	200	270
3	0.00036	−0.00014	−0.00058	−0.00031	−0.00014	0.00011	−0.00008	−0.00005	0.00001
7		−0.00043	−0.00020	−0.00029	−0.00003	0.00016	−0.00018	0.00016	0.00083
14			−0.00093	−0.00011	−0.00063	−0.00019	−0.00043	−0.00030	0.00014
21				−0.00011	−0.00004	0.00016	−0.00005	0.00037	0.00007
30					−0.00036	−0.00001	0.00005	0.00028	0.00016
60						−0.00023	−0.00032	0.00090	0.00041
100							−0.00009	0.00021	0.00020
150								0.00079	**0.00099**
200									0.00032

Table A.12 Coefficients of all possible trading rules: 100 % gray

	7	14	21	30	60	100	150	200	270
3	−0.00019	−0.00046	0.00022	−0.00015	0.00012	−0.00011	−0.00004	−0.00024	−0.00035
7		−0.00014	−0.00004	−0.00032	−0.00022	−0.00036	−0.00011	0.00004	0.00029
14			0.00012	−0.00036	−0.00020	−0.00005	0.00024	−0.00014	0.00022
21				0.00013	0.00061	0.00067	0.00071	0.00095	**0.00136**
30					0.00064	0.00032	0.00086	0.00054	0.00088
60						0.00104	0.00037	0.00013	0.00065
100							−0.00026	−0.00030	−0.00036
150								0.00045	0.00112
200									0.00021